FLYING START

FLYING START

A Fighter Pilot's War Years

Hugh Dundas

STANLEY PAUL

LONDON · SYDNEY · AUCKLAND · JOHANNESBURG

Stanley Paul and Co. Ltd
An imprint of Century Hutchinson
Brookmount House, 62–65 Chandos Place,
Covent Garden, London WC2N 4NW

Century Hutchinson Australia (Pty) Ltd
88–91 Albion Street, Surry Hills, NSW 2010

Century Hutchinson New Zealand Limited
191 Archers Road, PO Box 40–086, Glenfield, Auckland 10

Century Hutchinson South Africa (Pty) Ltd
PO Box 337, Bergvlei 2012, South Africa

First published 1988

Set in 10/12 pt Linotron Sabon by Deltatype, Ellesmere Port
Printed and bound in Great Britain by Butler & Tanner Ltd, Frome and London

British Library Cataloguing in Publication Data
Dundas, Hugh, *1920–*
   Flying start : a fighter pilot's war years.
   1. Great Britain.   Royal Air Force.   Pilots.
   Dundas, Hugh, *1920–*
   I. Title
   940. 54'42

ISBN 0 09 173732 X

Contents

Chronology

22 July 1920	Born, Barnborough Hall, near Doncaster.
July 1938	Left school.
End May 1939	Joined 616 (South Yorkshire) Squadron, Auxiliary Air Force, Doncaster, as Pupil Pilot.
2 June 1939	Began flying training in Avro Tutor.
Early June 1939	Commissioned as Acting Pilot Officer.
1 September 1939	Transferred to Service Flight.
3 September 1939	War declared.
2 October 1939	Awarded RAF Flying Badge of Wings.
10 October 1939	Posted to No. 2 Flying Training School, Brize Norton, for Advanced Training course.
3 March 1940	Rejoined 616 Squadron at Leconfield.
13 March 1940	First flight in Spitfire.
27 May 1940	Squadron to Rochford (Southend) to cover Dunkirk evacuation.
28 May 1940	My first engagement with enemy aircraft.
6 June 1940	Squadron returned to Leconfield.
3 July 1940	First score – share of one enemy bomber destroyed and one damaged over east coast convoy.
15 August 1940	'Adlertag' – squadron intercepts large force of German bombers approaching east coast near Bridlington. Personal claim – 1½ JU 88's destroyed.
19 August 1940	Squadron moved to Kenley, Surrey.
22 August 1940	Shot down by Me 109, baled out.
13 September 1940	Rejoined 616 Squadron at Kirton-in-Lindsay, Lincolnshire.

19–28 September	Squadron to Duxford sector daily to form part of Squadron Leader Bader's 12 Group 'big wing'.
28 November 1940	John Dundas killed in action.

1941

26 February	616 Squadron to Tangmere, Sussex.
8 May	Shot down by Me 109 south of Dover – crash-landed at Hawkinge.
July	Promoted Flight Lieutenant, commanding A Flight, 616 Squadron.
19 July	Awarded Distinguished Flying Cross.
8 August	Douglas Bader missing after dog-fight near Lille.
21 September	Posted to No. 59 Operational Training Unit, Crosby-on-Eden as a Flight Commander.
19 October	Posted to 610 Squadron, Leconfield.
20 December	Posted to command 56 Squadron, Duxford; promoted Squadron Leader. Squadron re-equipping with first Typhoon aircraft.

1942

30 March	Squadron moved to Snailwell (Newmarket).
30 May	Carried out first operational patrol by Typhoon aircraft while squadron temporarily based at Westhampnett (Tangmere sector).
25 August	Squadron moved to Matlaske (Coltishall sector).
5 November	Badly damaged by flak off Dutch coast, crash-landed Coltishall.
Mid-November	Posted to Duxford to form and lead first Typhoon fighter-bomber Wing; promoted Acting Wing Commander.

1943

January	Posted at short notice to Tunisia for duty as Wing Leader.
February-May	Combined duties as supernumerary Wing Leader 324 Wing (Spitfires) and GTI Fighters 242 Group, Souk-el-Khemis, Tunisia.

11 June	Appointed acting wing leader 324 Wing which moved to Malta for operations leading to invasion of Sicily.
10 July	Invasion of Sicily.
11 July	324 Wing landed in Sicily.
July	Appointment as Wing Leader 324 Wing confirmed.
3 September	Invasion of Italy. Wing moved to north coast of Sicily to cover Salerno landings.
9 September	Wing landed on Salerno beach-head.
10 October	Wing moved to Capodichino, Naples.

1944

Early January	Posted to RAF Staff College, Haifa. To Egypt on leave.
Mid-January	Staff College posting cancelled – recalled to Italy to serve on AVM Broadhurst's staff at AHQ Desert Air Force in place of officer killed in flying accident.
3 March	Awarded DSO.
31 May	Posted to 244 Wing as Wing Leader (five squadrons of Spitfires).
Mid-June	244 Wing converted to fighter-bomber duties.
10 July	Force-landed following damage by flak.
July–September	244 Wing providing close support to Eighth Army during battle for Gothic Line and subsequent advance.
19 November	Promoted Group Captain to command 244 Wing.

1945

23 March	Awarded bar to DSO.
9 April	Eighth Army breaks out from winter line across River Senio; beginning of final campaign.
7 May	Operations ceased.

CHAPTER ONE

Boy's-eye View

The two Skuas of the Fleet Air Arm swooped by in the opposite direction, twisting and jinking. In the headlong manner of their flight there was something reminiscent of agitated sheep running from dogs. And killer dogs indeed were at their heels. I saw the black crosses and the swastikas, plain and clear, and recognized them as Messerschmitt 109s.

Fascinated, I craned my neck to watch the five planes, now diving away behind and to starboard. From the leading Messerschmitt came thin trails of grey smoke as the pilot fired his guns. The group faded into specks which, in an instant, disappeared beneath the thick black smoke cloud rising from Dunkirk and stretching down the Channel for seventy or eighty miles.

Perhaps this little cameo lasted before my eyes for about five seconds, it was a lightning personal introduction to the use of guns in earnest and to the terrifying quality of air fighting. But I did not at that time have so much as one second to reflect upon it, for I was suddenly aware that the formation in which I was flying as last man in the last section was breaking up in violent manoeuvre. My own section leader, George Moberley, wheeled round in a climbing turn. As I followed I heard a confusion of excited voices on the radio. Then I saw another Messerschmitt, curving round. It had a bright yellow nose. Again I saw the ripples of grey smoke breaking away from it and the lights were winking and flashing from the propeller hub and engine cowling. Red blobs arced lazily through the air between us, accelerating dramatically as they approached and streaked close by, across my wing.

With sudden, sickening, stupid fear I realized that I was being fired on and I pulled my Spitfire round hard, so that the blood was forced down from my head. The thick curtain of blackout blinded me for a moment and I felt the aircraft juddering on the brink of a stall. Straightening out, the curtain lifted and I saw a confusion of planes, diving and twisting. My eyes focused on two more Messerschmitts, flying in quite close formation, curving down towards me. Again I saw the ripple of smoke and the wink of lights; again I went into a blackout turn and again the bullets streaked harmlessly by.

At some stage in the next few seconds the silhouette of a Messerschmitt passed across my windscreen and I fired my guns in battle for the first time –

a full deflection shot which, I believe, was quite ineffectual.

I was close to panic in the bewilderment and hot fear of that first dog fight. Fortunately instinct drove me to keep turning and turning, twisting my neck all the time to look for the enemy behind. Certainly the consideration which was uppermost in my mind was the desire to stay alive.

'A sincere desire to engage the enemy' – that, Winston Churchill has written, was the criterion by which Lord Haig had judged his fellow soldiers. That, above all else, was the impulse which Churchill himself admired and demanded in fighting men. I found out that day, 28 May 1940, over Dunkirk, in my first close encounter with Britain's enemies, how hard it is to live up to that criterion. When it comes to the point, a sincere desire to stay alive is all too likely to get the upper hand. Certainly, that was the impulse which consumed me at that moment that day. And that was to be the impulse which I had to fight against, to try and try and try again to overcome, during the years which followed.

But there was no thought of right or wrong, courage or cowardice, in my mind as I sweated and swore my way through that first fight over Dunkirk. When, at last, I felt it safe to straighten out I was amazed to find that the sky which only a few moments before had been full of whirling, firing fighters was now quite empty. It was my first experience of this curious phenomenon, which continually amazed all fighter pilots. At one moment it was all you could do to avoid collision; the sky around you was streaked with tracer and the thin grey smoke-trails of firing machine guns and cannons. The next moment you were on your own. The *mêlée* had broken up as if by magic. The sky was empty except perhaps for a few distant specks.

It was then that panic took hold of me for the second time that day. Finding myself alone over the sea, a few miles north of Dunkirk, my training as well as my nerve deserted me. Instead of calmly thinking out the course which I should fly to reach the Thames estuary, I blindly set out in what I conceived to be roughly the right direction. After several minutes I could see nothing at all but the empty wastes of the North Sea – not a ship, not a boat, not even a seagull, no thread to connect me with the precious, longed-for land.

This need to get in touch with the land pressed in on me and drove out all calmness and good sense. I saw that I was flying almost due north and realized that this was wrong, but could not get a hold of myself sufficiently to work things out. I turned back the way I had come, cravenly thinking that I could at the worst crashland somewhere off Dunkirk and get home in a boat. At last I saw two destroyers steaming at full speed in line ahead, and beyond them in the haze I could see the flat coastline of France. The sight of the two ships restored me to some measure of sanity and self-control. I forced myself to work out the simple problem of navigation which sheer panic had prevented me from facing. After a couple of orbits I set course to the west and soon the cliffs of the North Foreland came up to meet me.

Soaked in sweat, I flew low across the estuary towards Southend pier. By the time I came in to land at Rochford, the little grass field behind Southend where the squadron had arrived the night before to take part in the Dunkirk evacuation, a sense of jubilation had replaced the cravenness of a few minutes earlier. I was transformed, Walter Mitty-like: now a debonair young fighter pilot, rising twenty, proud and delighted that he had fired his guns in a real dog-fight, even though he had not hit anything, sat in the cockpit which had so recently been occupied by a frightened child and taxied in to the dispersal point, where excited ground crew waited to hear the news of battle.

That is the inglorious story of my first brush with the enemy. It is all quite fresh in my mind, when I like to turn on the tap of memory. Across the years, the sight of the hunted Skuas, the winking lights as the Messerschmitts opened fire, the two destroyers which brought me back to my senses – these things I see clearly, like stills from an old film. And I can conjure up in my mind and in the pit of my stomach the nasty sickening feel and taste of my first real experience of fear. This unheroic introduction to war was very different from the way I had imagined it would be – and, indeed, was rather different in some points of detail from the way I described it next day in a letter to my mother, discovered among her papers some forty years later. I had been fascinated by the idea of war from an early age. At my preparatory school, Aysgarth, in North Yorkshire, I spent many spare-time winter hours sitting by a radiator in a corner of the library studying *The Times History of the Great War* and its sister volume about the Boer War. I knew every picture in those large and well-illustrated volumes. I gazed fascinated at the drawings and photographs of gallant Britons, engaging the enemy in every kind of situation, by land, sea and air. And in my day dreams I led a thousand forlorn hopes, died a hundred deaths in a manner which aroused the astonished admiration of the entire nation.

There was nothing in my home or family background to explain this curious preoccupation. My father had been nearly forty when the first war started and to the best of my knowledge had never worn a uniform in his life. Although there had been soldiers and sailors among my forebears most had been more commonly engaged in politics and government than in fighting wars. The family prints which hung in the dining room and hall and in my father's study were mostly of eighteenth- and nineteenth-century gentlemen with hardly a sword between them.

With my elder brother John, who was to be one of the top-scoring fighter pilots before he was shot down and killed in November 1940, and my three sisters, Elizabeth, Alice and Charmian, I was brought up in the West Riding of Yorkshire in an atmosphere still coloured by the twilight of the Victorian age. My father was not well off, but he had been born in the purple and many of our friends and relations in Yorkshire still lived in a style and grandeur which I think can only have been marginally diminished in the aftermath of the First World War. Our own home, rented from a friend of my father, who owned the village of Cawthorne and its surrounding farms and woods, was comparatively modest and economy was a word we were taught to respect. But all around us was the world of great wealth and huge possessions, of enormous houses and limitless staff. My father went out shooting, often two or three days a week, throughout the autumn and winter, and from an early age it was my special joy to be allowed to go with him. In the summer there were tennis parties, bathing parties, cricket matches and picnics.

Every Sunday, wherever we were, there was church. At some houses where we stayed there were prayers every day before breakfast. One of our aunts carried this business of family prayers to rather an extreme stage. In her later years she acquired, or caused to be constructed, a portable altar for use at the shooting lodge on the North Yorkshire moors where she spent August and September with her family. The journey from the lodge to the nearest church had become too much for her and so on Sundays she assembled her altar in the drawing room and there conducted a modified form of morning service, leading the singing in a high quavering voice to her own piano accompaniment.

It was a way of life which seemed immutable and perfectly correct in all respects. It was as far removed as anything could be from the concept and associations of war. And yet the flame of martial ambition flickered constantly within me. At a certain stage in my boyhood I announced that I would like to join the Royal Air Force. This idea received no encouragement from my parents. However, by the time I left Stowe, in the summer of 1938, the shadow of war dominated all our lives. And I had only one ambition, to join an Auxiliary Air Force squadron as a pupil pilot and to do so without delay.

In all the history of arms there can seldom have been a body of men more outwardly confident and pleased with themselves than the pilots of the Auxiliary Air Force. We wore big brass 'As' on the lapels of our tunics and no amount of official pressure during the war would persuade us to remove them. The regulars insisted that those As stood for 'Amateur Airmen', or even 'Argue and Answer back'. To us they were the symbols of our membership of a very special club. Without the squadrons of the Auxiliary Air Force the Battle of Britain could not have been won. For out of the forty-two operational fighter squadrons available to Air Chief Marshal Sir Hugh Dowding in July 1940, twelve were Auxiliary units – more than a quarter of the entire force.

The pilots of the Auxiliary Air Force were lawyers and farmers, stockbrokers and journalists; they were landowners and artisans, serious-minded accountants and unrepentant playboys. They had two things in common – a passion for flying and a fierce determination that anything the regulars could do, the Auxiliaries could do better. In order to implement this determination a very high standard of flying had to be achieved, as every auxiliary pilot secretly appreciated, in spite of the assumed contempt for regulars and all their ways. The auxiliary squadrons had been raised on a territorial basis. They were equipped with front-line aircraft, but most of them had training echelons so that pupil pilots could be put through their entire training within the unit. For this reason, the adjutant and assistant adjutant seconded to each squadron were invariably regular officers who had qualified as flying instructors at Central Flying School.

Each squadron had a compliment of Auxiliary Air Force ground crews and tradesmen, sufficient in number to service all the squadron's needs. But in addition there were about seventy-five Regular RAF ground staff, whose job it was not only to train the auxiliaries but also to keep the squadron aircraft in good working order throughout the week, when the auxiliaries were engaged in their normal civilian pursuits.

In every auxiliary squadron I ever knew there was an exceptional spirit of enthusiasm and *joie de vivre*. This auxiliary spirit had been born, curiously enough, in White's Club, during the twenties. It was fathered by a large and (judging by pictures I have seen of him) somewhat florid aristocrat, Lord Edward Grosvenor, the third son of the first Duke of Westminster. This extraordinary man put his stamp on the auxiliaries and his influence lasted long after his death, in 1929. The flame which he lit was still burning strongly when the auxiliaries rose up to do battle in 1939. At the time when Sir Hugh Trenchard, Chief of the Air Staff, conceived the idea of an Auxiliary Air Force, Lord Edward was, so I have been told, in the habit of presiding, when in London, over the big table in the back dining room at White's. The big table dominated the room; Lord Edward dominated the table. He was a man who liked to sit long over his coffee, brandy and cigars without moving from the place where he had lunched or dined. You may smoke at any time in the back room at White's. So there Lord Edward sat, long into the afternoon, longer into the night, talking with his friends.

His conversation, as often as not, was of flying which was his passion and joy in life. He had been one of the first Englishmen to own a plane – a Bleriot, with which he offered himself to the Royal Naval Air Service in 1914. He had flown throughout the First World War and in peacetime his voice had been persistently raised to demand a territorial air force to match the Territorial Army. So when the first auxiliary squadrons were formed in October 1925, Lord Edward raised and commanded the celebrated 601 County of London Squadron. He recruited his pilots in part from his old wartime aquaintances, in part from his friends at White's.

Simultaneously, No. 600 City of London Squadron was formed. Its commanding officer, the Right Hon. Edward Guest, was quite unlike Grosvenor in character and habit. He was a most serious-minded man, who had given all his life to public service. Already fifty-one-years-old when he formed 600 Squadron, he had first fought for his country on the White Nile and in South Africa at the turn of the century. After the First World War he turned to politics. He was more likely to be found discussing affairs of state in the smoking room of the House of Commons than sitting far into the night in the dining room of a West End club. The personalities of Guest and Grosvenor shaped their squadrons, which shared the same airfield at Hendon and set the pattern for the whole Auxiliary Air Force. Guest looked for solid, worthy and conventional qualities in his officers. Grosvenor wanted mercurial men around and he did not care in the least whether they were conventional.

As new auxiliary squadrons were set up in various parts of the country, they looked to Hendon for their example. And they saw at Hendon a group of enthusiasts, well-laced with young men of great wealth, giving every moment of their spare time to the squadrons. They saw the steady city gents recruited by Guest; they saw the mercurial characters who surrounded Grosvenor. They saw the sacred cows of the RAF being openly laughed at and ridiculed. They saw conventions of dress and discipline casually flouted – Grosvenor himself, judging by a full-length portrait of him which hung in 601 Squadron officers' mess until the squadron was disbanded in the mid-fifties, setting the pattern in this respect, departing altogether from the accepted practice of RAF dress by wearing baggy blue breeches and Newmarket boots under a tunic which was cut very full, like a riding coat. They learned, to their awed amazement, that the regular officers at Hendon were openly referred to by the officers of 601 and 600 Squadrons as 'coloured troops'. But they saw also that the Hendon squadrons put so much time and effort into their training that their flying hours matched those of most regular squadrons. The 'amateur airmen' competed on level terms with the professionals in air-firing and bombing contests and frequently came out on top. They played a full and equal part in national defence exercises. They were a star turn at the annual RAF flying pageant staged at their home base, Hendon.

In an atmosphere combining light-heartedness with an underlying determination to excel at operational flying, the 'auxiliary spirit' was born and developed. It flourished strongly as new units were raised in all parts of the country. In the summer of 1939, there were fourteen squadrons ready to take their place in Fighter Command at the outbreak of war.

Number 616 (South Yorkshire) Squadron was one of the last to be formed. It was an offshoot of 609 (West Riding) Squadron, which my cousin and godfather, Harald Peake, had raised in 1935 and commanded ever since and in which my brother John was already serving as a pilot. Originally Peake intended to command 616 himself, but his appointment to the newly-created post of Director of the Auxiliary Air Force at Air Ministry was made at that time and he handed over the new squadron to another 609 officer, the Earl of Lincoln – later Duke of Newcastle.

I would have liked most of all to join 609 Squadron, so that I could have been with my brother John, but Harald Peake wanted recruits for the Doncaster squadron and that is where I was sent. There was also some special advantage for me in joining a squadron based at Doncaster. I had hoped when I left Stowe in the summer of 1938, that I would go on to Oxford, preferably following John's footsteps at Christ Church. However, although his career there had been highly successful from an academic point of view – he had gone down two months before his twenty-first birthday with first class honours in modern history and an award which enabled him to spend a year divided between the Sorbonne and Heidelberg University – it had evidently involved rather more high living than my father, who had only recently settled the bills, found acceptable. My own entry into Oxford had therefore been made conditional on the winning of a scholarship; and although, having sat the exams the previous winter, Trinity College had more or less promised me an exhibition if I returned the following year, that was an option which we decided not to pursue.

My father – very sensibly, as I see with hindsight – had expressed the view that a professional qualification was likely to be as useful to me as a degree and accordingly arranged for me to be articled to our family solicitors, Messrs Newman and Bond. Knowing that Oxford was out and having nothing else to suggest I fell in with the idea, although about the only attraction from my point of view was that Newman and Bond had a branch office at Doncaster and I foresaw the possibility of basing myself there and thus achieving a position where I might concentrate on the study of flight rather than on the study of law. This concept eventually worked out very well, but not before many difficulties and frustrations had been overcome.

A totally unexpected obstacle to my entry into the Auxiliary Air Force arose when I took the preliminary medical examination. I was failed and I could not find out why. I was simply told that I could reapply for examination in two or three months time. I suppose that I have never again been as fit as I was at that time. But I took all possible steps to get fitter still. When I got home from the office I set off on long runs along the dark country lanes. I practised all the tests which for some mysterious reason formed part of the examination, such as holding my breath till I nearly burst and standing on one leg with my eyes shut. I drastically restricted my cigarette smoking. And at the weekends I took an enormous amount of exercise.

All these efforts got me nowhere. Twice more I failed the preliminary medical examination. By April 1939 I seemed further than ever from my ambition of getting into the sky. I was seriously considering abandoning the whole idea and joining the Yorkshire Dragoons instead. If I could not get airborne I might as well get horseborne and gallop around in the regiment which was officered by many of my friends and some of my relations. However, at this point, someone pulled some strings. I was told that I would be seen by the senior medical officer from 12 Group Headquarters, who was going to visit Doncaster. His name was O'Malley – a tough, square, Irishman who had played Rugby football for his country. He was one of the few 'flying doctors' in the service, being a fully-qualified service pilot. He talked to me for a bit, gave me the most perfunctory examination and pronounced me fit. He said that no further examination would be necessary.

I was in.

I did not have any further general medical examination of any kind at the hands of the RAF doctors until after the war and I never discovered what was supposed to be wrong with me.

7

On several occasions during the war I ran into O'Malley, both in England and in the Mediterranean theatre. I never failed to remind him of my everlasting gratitude to him. He rose to high rank in the medical branch of the service and it always seemed to me entirely appropriate that he should have done so.

Thanks to O'Malley I was enlisted as a Pupil Pilot with 616 Squadron at the end of May 1939. I hurriedly arranged to transfer myself to the Doncaster office of Newman and Bond, and in order to be as close as possible to the scene of operations I went to stay at Bawtry Hall, the home of my father's elder sister, Aunt Evie, and her husband, Uncle Bertie Peake.

The first of a series of unforgettably exciting summers had begun.

CHAPTER TWO

Longing to Fly

During those golden days of high summer, before the old-fashioned world of my upbringing finally fell apart, I led a curious double life. As well as my room at Bawtry, I also had a room in the squadron mess at Doncaster. I could stay there as much as I liked, comfortably looked after by a civilian batman and live in company with other members of the squadron who, as war drew closer, made it their home. The CO, Lord Lincoln, was there most of the time. So was Teddy St Aubyn, a Lincolnshire landowner who had taken to the air after resigning his commission in the Grenadier Guards, following his marriage to Nancy Merrick, whose mother, 'Ma' Merrick, owned and ran one of London's most notorious night clubs. So was Dick Hellyer, the son of a Hull trawler owner and Dudley Radford and John Glover, the Adjutant and Assistant Adjutant. Others came and went.

It would be impossible to imagine a greater contrast than that which existed between my two homes that summer. Little had changed at Bawtry, I imagine, since my aunt had married and set up house there with Uncle Bertie in the mid-nineties. The interior of the house and the way of life were pure Victorian. In the front hall stood my aunt's tricycle, a big basket suspended in front of the handlebars; there it had always stood since the days when I was first old enough to notice anything, and no doubt since long before then, though I never once saw her use it. Beyond that the big inner hall was dark, cool and quiet, always heavy with the smell of flowers. Beside the broad main staircase stood the gong which rang at 8.45 a.m. for prayers; 9 o'clock for breakfast; 1.15 for lunch; 7.15 p.m., to warn that it was time to change for dinner, and rang for the last time when dinner was ready at eight.

Somewhere at the top of the stairs, hovering around the balustrade, would be, as like as not, the fluttering figure of 'Pickie' – my aunt's old lady's maid and dresser. 'Pickie' always fluttered. But she loved us children. She loved to talk and be talked to, to have a little joke and a laugh, for ever glancing anxiously over her shoulder, on the look-out for heaven knows what nameless danger. We loved her in return, and teased her, and lay in wait for her as she fluttered and rustled down the corridors, and jumped out at her so that she screamed and then screamed again at the horror of having screamed.

There was – or seems to have been, in retrospect – a curious superficial likeness between my aunt and 'Pickie', both small and frail and wispy, both covered from neck to ankle, day and night, in flowing clothes. But my aunt most certainly did not flutter, nor was she given to nervousness or screams. She moved resolutely about the house, steadily to-ing and fro-ing between the conservatory and the principal downstairs rooms. On her arm, very often, was a big shallow gardening basket, or in her hand a long-spouted watering can with which she refreshed the bowls and bowers of flowers which stood in every room.

Beyond the drawing room and my uncle's study, the French windows opened onto the long sloping avenue which led down to the lake and the boat-house, for many years the centre of attraction for all the children who stayed at Bawtry. The lawns were still mown in the old-fashioned way, with a big machine pulled by a pony which wore leather boots. At intervals during the day my uncle, who had been almost completely blind for many years, marched up and down the long broad gravel paths which ran along the front of the house between the lawns and the rose beds. Erect and firm of tread, always dressed impeccably in the Edwardian fashion, his linen starched, pearl-covered spats over shining boots, he promenaded briskly, holding the arm of his reader, a young lady whose job it was to read him the newspapers and his correspondence, and stopping from time to time to poke or point with his stick and ask some question.

The routine was strict at Bawtry and you were expected to stick to it. There was no departure from high standards in all things. When, one night, my youngest cousin, Edward, raided the cellar and split a bottle of champagne with me in the smoking room after the others had gone to bed he got a stern drubbing for leading the young astray. Alcohol, at Bawtry, was something to be consumed in proper quantities at meal times. Gentlemen, beyond a certain age, might have a whisky and soda instead of tea. Champagne in the smoking room at ten o'clock in the evening was clearly outside the accepted and acceptable code of behaviour. I loved Bawtry. I loved my aunt and uncle, who were unendingly and unfailingly kind and generous to me. I did not know, during those bright summer weeks, that I was living through the twilight days of their house and home, and indeed of their lives; that Bawtry which had been an important part of the lives of all the family, would soon cease to exist as we had known it and loved it and sometimes laughed at it.

Not knowing, I spent more and more time on the airfield at Doncaster. There I was emancipated as I started living, for the first time, with grown-up men quite unconnected with my family or its immediate circle. Some of them were in the habit of living their lives at a brisk pace. It was an exciting novelty for me to be able to press a bell at any time and order a whisky and soda or a cocktail. This was an operation performed at regular intervals by my new associates and, naturally, I felt that I should keep pace.

It was, incidentally, in the course of that simple operation that I became labelled with a nickname – 'Cocky' – which, so far as my years in the Royal

Air Force were concerned, entirely replaced and ousted my proper name. It was very soon after I had joined the Squadron and I was sitting by the fireplace in the mess one evening before dinner. On the wall at my side was the bell button. Teddy St Aubyn and others were there. Teddy felt the need for further refreshment and decided that I was conveniently placed to summon the mess steward.

'Hey, you,' he said, pointing to me, 'hey, you – Cocky – press the bell.'

I promptly did his bidding. But why had he described me as 'Cocky'? What had I done? Nervously, I asked him.

'Because I couldn't remember your name and because you look like a bloody great Rhode Island Red,' was the reply. And that was it. 'Cocky' I became in the Squadron. 'Cocky' I subsequently was to all who knew me in the Royal Air Force and to the majority of the new friends I made during the war years.

Over those evening drinks in the mess, I listened with fascination to my companions' talk about their many escapades and parties and, particularly, about the women involved in them. Teddy St Aubyn had an endless stream of reminiscences about life as a Guards subaltern in London. My experience of women at that time was absolutely zero and I listened to these hair-raising anecdotes with awe and amazement, which I tried hard to disguise under a knowing exterior.

But though this sudden emancipation, the pleasure and the shock of it, remains a vivid and important memory of that summer, all else was overshadowed by the excitement of flying. Because I was able to spend so much time at the airfield, during the week as well as the weekends, I got in more hours in the air than the other pupil pilots. The Avro Tutor bi-planes in which I did my dual instruction and early solo flying were, even then, relics of a bygone age in aviation. They were, in effect, modified versions of a plane which had been designed and built in the First World War. I must have been one of the last people in the RAF to be taught to fly in one of these antiquated but delightful machines.

For the last three weeks of June and throughout July I spent as much time as I possibly could in the air. I pestered Dudley Radford, the Regular RAF Flight Lieutenant who was attached to the squadron in the dual role of Adjutant and Chief Flying Instructor, to take me up or to send me up. I hung around the tarmac waiting for him to get through all his other duties and whenever I saw him, I pounced. He was a fine instructor, very firm, very fair. He got through to me at an early stage the need to combine delicacy with strength and decisiveness in handling an aircraft. Everything he did in the air was done with a sense of deliberation and firm purpose, but yet with a gentle hand. He used the same technique in handling his pupils. He was tough, but never rough. His utterances through the primitive intercom which linked instructor and pupil were often direct and uncompromising, to the point of being brutal if one did something really stupid. But he would be infinitely patient in going through a difficult manoeuvre again and again until it was mastered.

I remember particularly the slow rolls. I quickly mastered loops and stall turns. But slow rolls I hated and had great difficulty in achieving. I felt quite helpless when the machine was upside-down and I was hanging on my straps, dust and grit from the bottom of the cockpit falling around me. Again and again, when inverted, I instinctively pulled the stick back, instead of pushing it forward and so fell out of the roll in a tearing dive. Again and again Radford went through it, patiently and firmly, until at last I got it right.

In the middle of August the squadron went to Manston, a big grass airfield on the eastern tip of Kent, for annual camp. Those were magic days, the very last of peace, and I recall them clearly and nostalgically. The whole expedition was, in any case, an occasion of tremendous excitement for me. It seems almost unbelievable, looking back, that I had never even seen the English Channel before then and the south coast was quite unknown to me. The very fact of being able to climb up to a few hundred feet and thence to look across to France was a thrill. That was 'abroad' where I had never been. There were Calais and Boulogne, where the steamers were met by luxurious Pullman trains which carried people off on exciting journeys to exotic places. Just seeing them – and above all seeing them from the pilot's cockpit of an aeroplane – provided me with a vicarious sense of pleasure and seemed to bring me closer to the world I had read about and heard about but had not known.

We lived in tents at Manston, and in the mornings we got up at dawn and started flying when the sun was low and the mist still lay in the valleys. I was to be up at dawn more often than not in the course of the next six years – until I hated the sound of the word. But I learned then, with wonder and delight, the magic of taking off into the sky when the air was crisp and new and the young day sparkled. It was a magic which never failed to uplift me a little, even in the darkest days to come, when fatigue and fear were overpowering.

At lunchtime the day's work was over. The weather was fine and warm and we sat outside the mess tent drinking our gins or pints of beer. We usually sat there for quite a long time, as there was the whole afternoon for dozing or doing as we pleased.

We received a visit at Manston from the Air Officer Commanding No.12 Fighter Group, Air Vice Marshal Trafford Leigh-Mallory – a man who was to play an important part in our destinies in the months to come. The words 'AOC's inspection' were on the lips of every senior NCO in the squadron for forty-eight hours before the visit, but I am glad to say that so far as the pilots were concerned no time was wasted on any special spit and polish. I cannot remember any parade of any kind ever being carried out in 616 Squadron and the rule was not broken for Leigh-Mallory.

The dinner party which we gave for the AOC in our mess tent that night was a memorable affair. One of our after-dinner games involved climbing up the pole in the middle of the tent, squeezing out through one of the ventilation flaps on the side near the top, clambering over the ridge-pole, getting in again through the ventilation flap on the other side and sliding back to earth down the pole to the place you had started from. The most difficult part of this maneouvre, once you had achieved the initial climb, was getting out at the top, as the vent was divided into four small squares by two strips of webbing, one vertical and one horizontal.

After we had drunk a good deal of port on the night of Leigh-Mallory's visit it was decided that we should play the game. Two or three of us went up and down. Then someone suggested that the AOC should have a go. Very sportingly, he agreed. But he was not really built for that kind of thing. In the course of the passing years his figure had thickened. He got up the pole all right. But he had a terrible job squeezing out through the ventilation flap. We stood below and cheered him on. At last he plopped through and his face, purple with exertion, disappeared out into the night. The tent swayed and the ridge-pole sagged as he struggled across the top. His legs reappeared on the other side. He got half-way and stuck.

Shouting with laughter, we urged him on and his legs and buttocks wiggled and waggled as he fought his way through that canvas flap. Someone shinned up the pole and helped him with a few hearty tugs. He came out like a champagne cork, grabbed desperately at the pole and descended from a height of about ten feet in a free fall. Fortunately he was unhurt. He accepted a very large, very dark whisky and soda and left us hurriedly before we started playing something else.

A day or two after his visit we knew that we would not be going back to our civilian jobs when camp came to an end. I remember the exact moment when this fact became clear to me. We had been flying all morning, as usual, and were just beginning lunch. Someone came into the mess tent and announced the news of the infamous pact between Russia and Germany. Teddy St Aubyn, who was sitting opposite me, put down his soup spoon and said in a loud clear voice: 'Well that's .....d it. That's the start of the ......g war.'

I cannot say why this forcible pronouncement should have caused my mind to click over so positively to the realization that the moment had come at last. Teddy, after all, was not noted as a political pundit or a serious student of international affairs. But I heard his words and knew they were true.

Curiously enough they had an equally dramatic effect on at least one other person. The dignified steward provided by the catering company to run our mess while we were in camp was entering with a tray at that moment. He stopped dead in his tracks. He turned his head to listen to the man who brought the news. He turned it the other way to listen to Teddy's succinct interpretation of that news. Simultaneously his mouth fell open and the tray dropped from his hands.

We all went down to Ramsgate to swim that afternoon. It was a subdued bathing party and when we got back to the airfield we were told that the Auxiliary Air Force was being embodied into the Royal Air Force. We were to fly back to Doncaster the next day.

I thought: 'Well – that's the end of the old solicitor's office for me.' Then I thought: 'Now I shall be comparatively rich.' You might say, with justice, that these were not very profound or admirable sentiments to spring foremost to the mind of a young man about to go off to war. But at least my idea of comparative wealth in those days was a modest one. I had received an allowance of £80 a year from my father, paid in quarterly instalments. In addition I received a few shillings a week from the Auxiliary Air Force for time spent on training. Now, as an acting pilot officer, I would get eleven shillings a day. That, certainly, did not consititute great wealth; but it represented an increase of over a hundred per cent in my income.

That night we went off in a body to Margate's 'Dreamland' and whirled around on everything which that notorious fairground had to offer. I had an anxious few moments on a particularly vicious device called an Octopus. There was an excess of beer and spirits inside me and I thought that I was going to be air-sick for the first time in my life. Next morning, still queasy, we completed preparations for our return to Doncaster. I was ordered to fly one of the Tutors and was very pleased with myself because I had another pupil as a passenger, rather than the other way round. We flew in a formation of three aircraft, led by one of the fully-qualified squadron pilots.

At Duxford, near Cambridge, we came down for fuel. The Spitfires of 19 Squadron were lined up along the airfield boundary, well removed from the tarmac and hangars. Their pilots and ground crew sat on the grass beside them. It was my first view of a fighter squadron at readiness for action and I felt a strong sense of envy for those seemingly God-like pilots with their beautiful machines, whose graceful lines I was then seeing for the first time.

The excitement of the scene was enhanced by the fact that a machine-gun post, manned by steel-helmeted aircraftmen, had been set up in the middle of the airfield. This can hardly have increased the operational efficiency of the field, but it added impressively to the warlike atmosphere of the place. We reached Doncaster late that evening. I had not yet started formation training, but I tucked my Tutor's wing as close as I dared to the leading plane as we dived over the airfield. I was childishly happy, after we had landed, to find the performance had been observed with approval by the experienced pilots of the Service Flight, who had flown back in their Gauntlet fighters and landed before us.

During the final days of camp at Manston I had started my graduation from initial trainers to 'service' planes. There was a dual-control Hawker Hind in the squadron which was used for this purpose. It was stretching a point a bit to describe it as a 'service' plane; it was in fact an obsolete two-seater biplane light bomber, with a top speed of about 160 mph. However, it seemed to me at the time to be a most formidable machine. Its engine, of course, was very much more powerful than that of the Tutor. Flying it, after the Tutor, was roughly equivalent to finding oneself in charge of a vintage Bentley after pottering around in an old Austin tourer.

I did not know it at the time, but it seems that a decision was made by the

CO and flight commanders to push my training ahead at top speed, so that I could stay with the squadron. And so, early in September, there came the magic moment when I was let loose for the first time in a Gauntlet.

It is alarming, in retrospect, to consider that any fighter squadron should still at the beginning of the war have been equipped with such old-fashioned machines. But to my eyes the Gauntlet epitomized everything a fighter plane should be. The stubby little radial-engined bi-plane, open cockpit set midway between nose and tail, seemed tailormade for Biggles – or Errol Flynn. We taxied them down to the gun-pits, jacked them up into the flying position, started the engines and fired the guns through the propellers to make sure that their synchronization system was working properly. The rat-tat-tat of the guns, combined with the clatter of the engines, produced exactly the right kind of noise to excite the senses of anyone who had seen and been deeply stirred by the epic films of air fighting in the First World War, such as *Hell's Angels* and *Dawn Patrol*.

The realization that our Gauntlets were not, after all, entirely suited to the needs of the time was brought sharply home to me two or three days after I had made my first flight in one. There I was, climbing away from the airfield, goggles down, scarf neatly adjusted, when a twin-engined Hampden, belonging to a bomber squadron based at nearby Finningley, came alongside and passed me on the climb. The pilot made a rude and familiar sign as he soared contemptuously past me. I opened the throttle wide, but could not keep up. I then understood why the older and wiser pilots in the squadron were so passionately anxious for our delightful little Gauntlets to be replaced by Spitfires or Hurricanes.

Around the second week of September I was given my Wings. That was a moment of tremendous triumph and satisfaction. It was also, I suspect, an act of low cunning on the part of my superiors in the squadron. The winning of Wings had always been regarded in the RAF as something which was not to be taken lightly; there were stiff written examinations as well as comprehensive tests in the air, usually carried out by a visiting Central Flying School instructor. Wings, after all, were the hallmark of professional competence.

In my case the hallmark was stamped on pretty base metal. I took no examination of any kind. I performed in the air for no visiting instructor. I was just told to get a pair of Wings and to have them sewn onto my tunic. This I did in record time. I expect that I also spent a record amount of time admiring the effect in my looking-glass. The explanation for this unorthodox performance was to be found in an instruction to all auxiliary squadrons' commanding officers. It decreed that pilots who had not yet attained Wings standard were to be posted away from the squadrons to flying training schools.

All through that crisp, suspenseful September we flew and flew and flew. Every day, two or three times a day, I was sent up to cram into a few weeks an amount of training which normally stretched over several months. First I was taught to fly in tight formation, wing-tip tucked in between my leader's wing and tail, so that with a medium-length trout rod I could have reached out from my cockpit and tapped him on the top of his leather-helmeted head. Then I was introduced to the horrors of staying in formation in cloud, sometimes so thick that if you drifted two wing-spans apart the plane you were following disappeared from sight in the swirling murk. And when you stayed there, tucked in tight, you had to fight back the feeling that your leader was carrying out extraordinary and alarming manoeuvres – climbing vertically and about to stall, or falling away in a steep spiral– when in fact he was flying absolutely straight and level.

Sometimes I was sent up alone, to practice aerobatics. And on one or two occasions I fell for the temptation which has too often ended in disaster for foolish, flamboyant, inexperienced but over-confident young pilots. I flew off to the house where my best girlfriend was staying and looped and rolled and span and generally beat the place up. She was impressed, I am glad to say, though her sister was not – and said so. However, her sister subsequently married a bomber pilot, which explains the difference in outlook.

During September, Pelham Lincoln, our CO, was posted away. He was considered too inexperienced as a pilot to lead a fighter squadron in war. His place was taken by a very regular officer named Beisiegel, inevitably known to one and all as 'Bike'. Other regular officers were posted to the squadron to bring the pilot strength up to establishment. We quickly inculcated them with the auxiliary spirit and there was no division between us. In the months to come, indeed, it seemed that they, no less than the rest of us, were original members of the squadron.

Some time in September we were told that our Gauntlets were to be replaced by Hawker Hurricanes and one day a single specimen of this famous breed was delivered to the squadron. We looked it over with awe and some apprehension. We climbed onto the wing and gazed with amazement at the mass of knobs, switches and instruments in the little cockpit. Eventually Ronald Kellett, a stockbroker by profession, who had formerly served with No. 600 (City of London) Auxiliary Air Force Squadron and now commanded 'A' Flight, announced that he would fly it. We gave him plenty of good advice and assured him that we would all be standing by to watch the accident, which fortunately did not occur.

However, our excitement was shortlived. Next day the Hurricane was taken away again. We were told that we were to get Spitfires instead – a development which should have made us all very happy and which I certainly can look back on as a turning point in my life. But of course I did not then know that I would soon be entering into a loving and adventurous relationship with the Spitfire which was to last for nearly ten years.

And so the first months of the war passed by in a happy blur of flying by day and parties by night. Once or twice I got home to Cawthorne for a night or two. On one occasion my father and I, with my brother-in-law, Johnnie Muir, who was on leave from India, spent a happy day on a neighbour's estate shooting partridges. On the surface, nothing had changed in the ordinary world from which I had so abruptly been removed. But time and life were in suspense. With a little bit of luck, it seemed, we might yet get our Spitfires before the storm broke, and so be in on the fighting from the start. So far as I can remember I was completely happy. In fact I think that perhaps I had never been happier. It really did seem a great deal better than the prospect of another four years as an articled clerk in a solicitor's office.

Then the bubble burst. The staff officers at Group Headquarters had been doing a little research. They stumbled upon the fact that although Acting Pilot Officer Hugh Dundas had been awarded his Wings, he still had less than a hundred hours flying experience. It was decreed that I must go on an advanced training course at Brize Norton, a few miles west of Oxford. Efforts were made to save me from this ignominious fate, which had already overtaken several others who had joined the squadron at about the same time as I had, but authority was adamant. Sadly I packed my bags and departed, leaving my thirteen year-old Austin Seven as a keepsake for the squadron. I knew it would not get as far as Brize Norton, anyway.

I vowed I would be back at Christmas.

CHAPTER THREE

Spitfire

Because of an appendix operation my vow was unfulfilled and I finally rejoined the Squadron late in February. And although I think that I had gained nothing at all by going to Brize Norton, it was also true that I had lost nothing.

Like most units of Fighter Command, 616 Squadron still had not made contact with the enemy. It had moved to its war station at Leconfield, a few miles north of Hull, and re-equipment with Spitfires had just been completed when I got back. The squadron's main duty was to provide protective patrols, usually consisting of a 'section' of three aircraft, over the convoys which steamed constantly up and down the east coast. It was tedious work, often carried out in bad weather. Occasionally a second section would be scrambled to investigate an 'X-raid' – the code-word for a radar plot of an aircraft which was unidentified but not definitely known to be hostile. Tedious work it may have been, but at least it was 'operational'. It was about as close to actual war as you could get at that time, in Fighter Command. And I was agog to join in it.

On 13 March 1940 I made my first flight in a Spitfire.

The story of the Spitfire's origins and development has been often told. It is an epic story, strong on drama and pathos. R.J. Mitchell, the young man of genius who designed the plane itself, and Sir Henry ('Pa') Royce, the much older man of equal genius who designed the Merlin engine which powered it, both died before the plane was more than a fledgling. In the RAF itself, no one had more to do both with the selection and with the introduction of the Spitfire than Sir Hugh Dowding. As Air Member for Supply and Research from 1930 to 1936 he had been a leading and, perhaps decisive advocate of the specification which both the Spitfire and the Hurricane were designed to fill. And then as Commander-in-Chief of Fighter Command from 1936 to the end of 1940 he supervised the extraordinarily rapid evolution of a command structure and *modus operandi*, involving absolutely new technical devices such as radar, direction-finding procedures and effective aircraft-to-aircraft and ground-to-aircraft radio communication, which enabled the Spitfires and Hurricanes to be used with maximum effect. And so, unlike Mitchell and

Royce, Dowding lived to see the Spitfire story develop in its entirety. However his active participation, although crucial, was short-lived, as he was ignominiously pushed into the background and retired from active service very soon after the Battle of Britain had been won, under his command, and thanks to the strategies and techniques which he had developed.

There is something Wagnerian about these facets of the Spitfire story, the more so since it is certainly true that there never was a plane so loved by pilots, combining as it did sensitive yet docile handling characteristics with deadly qualities as a fighting machine. Lovely to look at, delightful to fly, the Spitfire became the pride and joy of thousands of young men from practically every country in what, then, constituted the free world. Americans raved about her and wanted to have her; Poles were seduced by her; men from the old Dominions crossed the world and the oceans to be with her; the Free French undoubtedly wrote love songs about her. And the Germans were envious of her.

Little did I know as I taxied in from that first Spitfire flight that I would not taxi in from my last until late in 1949. We went through the war together, with only a year's separation when, in 1942, I temporarily – and not very happily – flirted with the Typhoon. We played a part in the re-establishment of the Auxiliary Air Force after the war. Indeed it was only a love affair of a more conventional nature which brought it all to an end when, towards the end of 1949, I decided that in view of my forthcoming marriage my week-ends would be better spent at home than with 601 squadron which I was then commanding.

In all those years no misfortune which came our way was ever the fault of the Spitfire. Owing to the loss of my second log book – the first ran up to the end of July 1942 – I do not know exactly how many hundreds of hours I spent in a Spitfire's cockpit, over sea, desert and mountains, in storm and sunshine, in conditions of great heat and great cold, by day and by night, on the deadly business of war and in the pursuit of pleasure. I do know that the Spitfire never let me down and that on the occasions when we got into trouble together the fault was invariably mine.

But, back to the beginning: after that first flight and a few hours gaining experience in handling the plane – aerobatics, formation flying, mock air-to-air attacks – the big day came when I was declared 'operational' and posted to 'B' Flight. This was commanded by Flight Lieutenant Denis Gillam, a regular officer who had joined the squadron soon after the beginning of the war. He allocated me to Green section, which was led by a fellow auxiliary, Flying Officer George Moberley.

Gillam was a professional with the highest standards as an RAF pilot, which he impressed upon us constantly and firmly. He wore the red and white striped ribbon of the Air Force Cross, a decoration awarded for gallantry in the air but not in the face of the enemy. He had earned it before the war when, as a member of the RAF Meteorological Flight based at Aldermaston, in Northern Ireland, he had been engaged in flying Gladiators, equipped with devices for measuring meteorological information, on daily climbs through any kind of weather to maximum altitude. It is an exercise which would not cause the slightest anxiety to the pilot of a modern plane, enjoying every kind of blind-flying and navigational aid. But for Gillam and his fellow-officers in the pre-war Met. Flight, who had to rely on the most rudimentary blind-flying panel – not even an artificial horizon – it was a highly-demanding and hazardous operation. That he had been well chosen for such a task was to be confirmed by his remarkable, indeed awesome, record as a fighting pilot over the next five years.

George Moberley, who had been transferred from 609 Squadron when 616 was established, was a tall and exceptionally good-looking man, several years older than myself, who became, all too briefly, a close and dear friend. He had served for a short time with the navy – with some kind of a reserve commission, I believe – and had a job with an engineering firm in East Anglia. His skill as a pilot, combined with his looks, breezy manner and his ownership of a Lagonda tourer and a fourteen-foot international class sailing dinghy, made him a heroic figure in my young and impressionable eyes.

The other pilots in the section had all been posted to the squadron since the beginning of the war. They were Flying Officer 'Scottie' Scott, a regular short-service commission officer; Sergeant Burnard, a regular non-commissioned pilot (Sergeant Pilots were very much to be respected and looked up to in the pre-war RAF); and Sergeant Hopewell, an amiable Yorkshireman, with an astonishing aptitude for minor transgression who was generally, and affectionately, addressed as 'Sergeant Hopeless'.

Throughout the spring of 1940 we waited impatiently for our war to start in earnest. Across the North Sea the battle raged in Norway, but affected us not at all. Then we read in our newspapers of the blitzkrieg breaking across north-west Europe. We heard of fierce fighting, with the Hurricane squadrons of the RAF hard-pressed but scoring heavily in their daily battles with German bombers and fighters. We knew that many squadrons of Hurricanes from Fighter Command were being sent to join the battle. Enviously, we learned that some of our own auxiliary pilots – men who had joined as pupils with me – had been posted to squadrons in France on completion of their stints at flying training schools. But still we patrolled our unglamorous convoys, or chased around the sky after X-raids which never turned into genuine 'bandits'. We practised our stereotyped 'Fighter Command attacks' and we carried out mock attacks against potential ground targets in our sector, such as the big power station at Ferrybridge, to provide practice for searchlight crews and anti-aircraft gunners.

It soon became evident to us that there was no prospect of the squadron being sent to France. Only Hurricanes were going. We would, no doubt, have felt less disappointed about that state of affairs if we could have known about and appreciated two things – the vital importance, so clearly discerned and insisted upon by our Commander-in-Chief, of holding the Spitfires back in Britain in order to maintain an effective air defence force; and the hopelessness of the brief but savage air battle taking place over France.

When the Battle of France was clearly lost and the evacuation from Dunkirk began, Dowding could no longer hope to keep his precious Spitfires out of the continental battle. He was ordered to throw in his full strength to protect the embarking armies. Even so he declined to concentrate all his squadrons in the south-eastern sector of the country which was within range of Dunkirk. It was the first manifestation of a rigid rule adhered to by Dowding throughout that dreadful summer − never, however great the emergency, however pressing the onslaught in any one area, to be persuaded to throw in all his squadrons from other parts of the country. He was not prepared to denude the midland and northern cities of their defences. That rule he adhered to even during the most testing periods of the Battle of Britain. That rule he now adhered to during the evacuation of Dunkirk. And so we continued to watch and wait in the fervent hope that our time for action would soon come around. And it did.

On the morning of 27 May I was sitting at breakfast with several other squadron pilots when Wing Commander L.G. Nixon, the station commander, bustled in, all pink and white with excitement.

'Well, chaps, this is it. You'll be in action by this evening. Lucky dogs!' And he proceeded to purse his lips and make a farting noise, apparently intended to simulate the sound of machine guns. He then, without further ado, ordered two poached eggs on toast, making special point of the fact that there should be lots of butter on the toast. I remember quite distinctly that my excited, tummy-tickling reaction to his unexpected announcement was interrupted by the thought that he was a greedy old devil. When he was quite satisfied that he had got across his point about the eggs, the toast and the butter he turned his attention to us again and satisfied our pressing demand for more information.

We were to go that afternoon to Rochford, he told us, to take over from No. 74 Squadron. We would be engaged in the Dunkirk operation. Rochford? Rochford? Where the hell was Rochford? Breakfast was forgotten by everyone except the station commander. He told us that Rochford was a small airfield outside Southend and came under command of the Hornchurch sector. Between mouthfuls of cereal he gave us a few more bursts of simulated machine gun fire, which was testing for the people sitting next to him. He also gave us the benefit of his advice derived from experience as a fighter pilot in the First World War. It boiled down to three words: 'Watch your tail.' If some of us had taken more notice of Wing Commander Nixon's wise words, it would have saved much pain and grief.

In spite of the unfeigned eagerness which I had felt to get into battle, the realization that before nightfall I might be involved with real bullets, and with all the dreadful consequences they could have on a fighter plane − and most specifically on its occupant − came as a jarring shock. Here, in Yorkshire, only forty miles away from home, I felt safe and normal. The 'cold war', the convoy patrols and dawn readiness, were an exciting and glamourized extension of flying training at Doncaster. Occasionally someone in the sector got hurt in an accident. Poor Tony Wilson had crashed into the sea and been drowned. Night flying in our Spitfires was alarming and seemed to me definitely unsafe. But at no stage had I come face to face with the more deadly dangers of actual air warfare.

Now I knew that I was about to do so. The knowledge was unsettling, but not much worse than the sort of feeling you get before going in to bat an important match against accurate fast bowling.

I was nervous. But I still did not know what it meant to be afraid.

We circled Rochford in the late afternoon. The grass airfield looked small – only about a thousand yards long in the east-west direction and much less across. A railway line ran along an embankment at the eastern end – an obstacle which complicated an approach from that direction and constituted a dangerous trap for anyone misjudging his landing and overshooting from the west. It was clearly impossible to land a Spitfire other than in one of those two directions.

There were two small hangars on the south side of the field, old-fashioned buildings of the type which might have been put up during the First World War. There was a club-house in the south-east corner. There were no barrack blocks, parade grounds or other paraphernalia of a normal RAF station. This was a field which had been used as a centre for private flying enthusiasts and was now pressed into war service.

Marcus Robinson, who had recently been posted from one of the Scottish Auxiliary Air Force squadrons to take over command of 616 from Beiseigel, urged great caution in landing as we broke up into sections. But in Green section we liked to do things with a swagger – particularly when battle-experienced pilots of a famous regular squadron like Number 74 would be watching. We held tight formation as George Moberley put us first into line astern and then, on the final approach, into echelon starboard. He brought us in slowly and carefully, with plenty of engine, and we touched down together with our wing-tips only a few yards apart. Still keeping close together we taxied towards the hangars, where the ground crews of 74 Squadron were waiting to receive us.

One thing is certain. The pilots of 74 who stood around outside the crew room were not in the mood to be impressed by our formation landing. They were far beyond the stage where anything much was going to impress them, except success in battle. Their appearance and demeanour came as an immediate shock and as a check to my carefree enthusiasm. They were dead tired and they looked it; they had suffered losses which had hurt and the fact was apparent; they were glad to be going north for a rest and they did not mind saying so. I have no doubt that they regarded us as a bunch of amateurs, soon to learn our lessons.

The effect was chilling. An understanding of what was coming was brought home to me and the mood of gaiety and excitement was replaced by a more nervous form of anticipation. But still the schoolboy approach to war was in the ascendant. There was something pleasingly and romantically appropriate about the whole situation – the little airfield with its old-fashioned hangars which might have been used as a set for a film, the strained faces of the men we were relieving, the urgent bustle of refuelling and preparing our Spitfires for take-off spurred on by a message from sector operations that the squadron might be required for a patrol that evening.

I have always had a curious way of remembering episodes and thoughts from the past in conjunction with some quite trivial association. I now remember as though I were looking at an old film most of which was blurred but which suddenly reflected a few frames with absolute clarity, a moment when I was alone that afternoon and what I was saying to myself. We had walked over to get some tea in the old club house, which was being used as an officers' mess. I was alone in the wash room. I talked out loud to my reflection in a mirror on the wall. 'Well, Hughie,' I said to myself, 'you couldn't insure your life now, for love nor money.' I said it several times over, because I thought it sounded rather dramatic.

Trite? Certainly. Trivial? Yes. But I think that cameo-memory is enlightening. It shows me mentally aware of the fact that two months before my twentieth birthday, sudden death was an imminent possibility. But it shows me facing this knowledge with a cliché, *Boy's Own Paper* style. Even in that grave moment I still saw Errol Flynn looking over my shoulder in the mirror.

Before going to tea we had watched the majority of 74 Squadron take off for Leconfield, led by a good-looking, square-jawed young Flight Lieutenant with straight fair hair and hard, unsmiling blue eyes. His name, we were told, was Adolph Malan. He was a South African and he was nick-named 'Sailor' because he had at one time served as a merchant seaman. He was pretty hot stuff as a fighter pilot, his colleagues told us, and had just been awarded the DFC. Pretty hot stuff? Later, of course, Sailor Malan was to prove himself outstanding among the half-dozen most brilliant fighter pilots of the war.

The CO of 74 Squadron, Squadron Leader White, stayed behind to tidy up his loose administrative ends. He came to the mess to have tea with us and to answer our questions. He was rather older than most men in the fighter squadrons, in fact, he seemed to me to be positively middle-aged – which meant that he might have been approaching thirty.

He was extremely lucky to be sitting there at all. He had been shot up during a patrol west of Dunkirk and had force-landed on the airfield at Calais, already surrounded and cut off by the German army. Anyone would have thought that for Squadron Leader White the best prospect available was a prisoner-of-war camp. The airfield was within range of enemy gun fire. There was little hope of getting to the coast and home by boat.

But Sailor Malan had different ideas. He carried out a rescue operation which became a classic in RAF history. He flew off in an unarmed, two-seater Miles Master trainer, escorted by a section of Spitfires, and landed at Calais.

As White climbed into the plane and they taxied out for take off, Messerschmitt 109s swooped in to attack and the escorting Spitfires plunged down to drive them off. Malan got the Master away amid a hail of fire and flew out to the coast, twisting and turning to evade destruction. He made it. White was brought home intact so that, thanks to Malan's gallantry, his experience as a highly trained regular officer was available to the service for the rest of the war.

Over tea we listened to White's account of the air battles fought by 74 Squadron, of their successes and of the losses they had sustained, of the confusion over Dunkirk resulting from the thick pall of smoke rising from burning oil installations and of the numerical superiority of the Luftwaffe. While he talked the order came through: 616 Squadron patrol Dunkirk, take off 18.30 hours.

No one who flew over Dunkirk during those momentous days will ever forget the sight. The black smoke rose from somewhere in the harbour area, thick, impenetrable, obscuring much of the town. As it rose it spread in patches, caught up in layers of haze and cloud. But still the greater part thrust upwards to a height of between twelve and fifteen thousand feet, where it was blown out in a lateral plume which stretched for many miles to westward, over Calais and beyond, down the Channel.

Beneath the smoke the air was thick and hazy, so that a pilot's-eye view of the scene was limited and confused. I think probably my inexperience made conditions seem more baffling than they really were. I dare say that if we had had a more experienced leader things would have seemed – and probably would have been – different. But my chief memory and impression of the Dunkirk patrols is of their nightmarish quality. We seldom seemed, somehow, to be in the right place at the right time. Again and again we would arrive over the beaches just after a Luftwaffe attack, or leave just before one. Other planes – sometimes hostile, sometimes friendly, sometimes un-identified – swirled around in the haze and smoke. Often we were engaged in a short, sharp action, usually resulting in the squadron becoming split up into sections or individual aircraft. On other occasions we returned to Rochford without having fired our guns.

On that first evening we saw nothing. The first shock of the battle came next day and I have already described its effect on me. Marcus Robinson was shot up, and landed at Manston. Sergeant Pilot Ridley, one of the regulars posted to the squadron after the outbreak of war, also came home with a badly damaged plane and a graze on the head from a bullet which had passed through his cockpit. Dick Hellyer did not return at all. Against these reverses the squadron claimed one Me109 destroyed by George Moberley, another one damaged by the same pilot and a third probably destroyed by Pilot Officer 'Scottie' Scott.

'One of our pilots is missing'. For the first time I contemplated the loss of a man I knew well, someone I had lived with closely over the months. Teddy St Aubyn was deputed to collect Dick's personal belongings and performed that melancholy task with typical panache, regaling us afterwards with details of some of the finds he had made. Happily, however, Hellyer turned up twenty-four hours later, having force-landed on the beach and hitch-hiked home by boat.

A couple of days later a large parcel arrived in our officers mess, addressed to Marcus Robinson. He seemed a bit coy about it, but our curiosity was insistent and he could not avoid opening it in front of us all. It contained a bullet-proof waistcoat, a revelation which gave rise to much hilarity. After weighing it in our hands we decided that it was not at all the kind of garment to be wearing at a time when a descent into the sea was always a strong possibility. Furthermore, it seemed a trifle inadequate when judged as a means of protection again 20mm cannon shells tipped with high explosive charges.

A more useful item of equipment, with which we all provided ourselves following a shopping expedition in Southend for that purpose, was a rear-view mirror. This was a refinement which had not been fitted to the Spitfire by the manufacturers. But experience in air fighting soon showed the need for it. We bought ordinary motor car mirrors and had them screwed onto the top of the windscreen. Throughout the rest of the war I would never willingly have flown operationally without one.

Towards the end of the Dunkirk evacuation we began to meet some of the soldiers who had been brought home from the beaches. Naturally, on the first of such encounters, we expected warm appreciation. It was a shock to be met by open hostility. 'Where the hell were you?' – that was the question we were often asked, in a tone of anger and contempt. It was no use saying that we were patrolling Dunkirk two or three times a day. It was no use pointing out that every Spitfire and Hurricane squadron in south-east England was putting maximum effort into covering the withdrawal, that casualties were heavy and victories numerous.

The answer, all too often, was the same: 'Well, our lot didn't see you. But we saw a hell of a lot of German planes doing just what they bloody well liked. The RAF must have preferred it out to sea.'

It was bewildering and distressing. Later the argument became bitter in the extreme, it marred the memory of an operation in which men of all three services performed with great valour and endurance. At the time – and for some time afterwards – I could not understand why so many soldiers on the beaches, and so many sailors also, believed implicitly that the RAF had not been there.

Looking back at it, the explanation seems quite simple. Firstly, the enemy had far more planes available to use over Dunkirk than we had. Secondly, those planes could be based closer to the scene of action than ours. And thirdly, the nightmarish, smoke-laden conditions, which I have already described worked to the advantage of the Germans. They were able to nip in and out round the curtains of smoke and smog and cloud, at varying altitudes and from varying directions, on bombing and strafing missions which might last only a few seconds but were none the less devastating to the defenceless men who formed their targets. Meanwhile the RAF might have two or three squadrons on patrol to cover the whole area, at all altitudes. Our own radar installations were too far away to provide us with any help in locating and intercepting the enemy planes. It was a game of blind man's buff and we were the ones with the blindfold.

In such circumstances it is not at all surprising that, again and again, our troops and our rescue vessels were bombed and strafed by German planes which we never saw, even though we were somewhere in the area, and that the victims of those attacks felt themselves to have been terribly let down by the RAF. It is not surprising; but at the time, with our own casualties mounting, it was hard to accept the derisive criticism to which we were subjected by the men whom we thought we had been helping to save.

Personal bitterness was not alleviated by the kind of experiences which were common in the squadrons. I recall two episodes affecting 616 Squadron. Jack Bell, the solicitor from Lincoln and an original member of the squadron, ended up in the sea after a fight in which he attacked a formation of Messerschmitts strafing a British ship. He shot one down before one of the others got him and he baled out from a low altitude. After a considerable period in the water, a Royal Navy open boat appeared close by. Bell shouted for help and the boat crew came over to investigate, when they saw that he was an airman they pushed him away with their oars and would have left him had not an officer on board insisted that he should be picked up. The explanation he received was that all fliers around those parts were naturally assumed to be German.

A few days later, when the evacuation was almost complete, we were called to readiness before dawn. Ground fog was thick on the airfield and Marcus Robinson reported to Sector Operations that a take-off was out of the question until visibility improved. But we were told that the final convoy of rescue ships was leaving Dunkirk and that it had to be protected at all costs. Every available squadron was to take off. There had been too many complaints about the RAF from the army and there was to be no question of leaving this last convoy unprotected. A few minutes before we were due to go, Robinson again protested. He was told curtly that an order had been received direct from 10 Downing Street to the effect that all units should take off, regardless of conditions.

We climbed into our cockpits and groped our way across the airfield, section by section, navigation lights switched on. I was on George Moberley's left in Green section. On his right was 'Scottie' Scott. George lowered his seat and kept his head down inside the cockpit, evidently deciding that this must be a purely instrument take-off. We tucked our wings in alongside and put ourselves into his hands. I could only see George's plane clearly, as we careered across the field through the fog. Scottie was a half-seen shadow to his right and he disappeared altogether just after take-off. As we popped out of the white blanket into the sparkling blue morning sky, two or three hundred feet above the airfield, Scottie was still not to be seen. Later we learned that he had crashed in the final stages of his take-off and had been killed.

It is not surprising that we were ourselves inclined to bitterness when it was suggested that the RAF made no effort to support and protect the evacuation.

But that was our last patrol over Dunkirk. Two days later, on 6 June, we returned to Leconfield and 74 Squadron flew back to its parent sector. If we had not done brilliantly, we had not done too badly either. The squadron claimed five enemy aircraft definitely destroyed, four probably destroyed and five damaged. We had lost one pilot; two of our Spitfires had been shot down and destroyed over Dunkirk, but the pilots had got home safely; two more aircraft had come back in a badly damaged condition.

As for myself, I had still to break my duck. But after the first action, when fear had got the upper hand, I had succeeded in remaining comparatively calm and controlled when enemy planes were about. I was desperately disappointed about not having been able to claim a victory. But I felt quite pleased with myself, all the same – like a schoolboy who has played, without scoring but without letting the side down, in an important and successful football match.

CHAPTER FOUR

The Waiting Weeks

High summer, 1940 – that is a time to look back upon with wonder, a time to have been alive and British. Above all, perhaps, it was a time to have been a fighter pilot in the Royal Air Force.

Over the years almost as many words must have been written about the conduct of the Battle of Britain as there were bullets fired in the course of the conflict. I shall not add to them, except to the extent of describing the background against which my own exceedingly inglorious part was played out.

Long afterwards, in September 1960, I took part in a television programme marking the twentieth anniversary of the Battle. A distinguished German fighter pilot was among those present. He expressed the view that no one had won the Battle. It had just come to an end. It had come to an end because the Germans had stopped fighting it. Another participant in the programme was our old Commander-in-Chief, Lord Dowding. Not unnaturally he took a different view. This is what he said: 'Let him say that the Germans stopped the battle, if that's what he wants. Very well, they did stop the battle. Certainly they stopped it. They stopped it because they had failed in their objective, which was to clear the RAF out of the skies so that they could invade this country. They stopped it because they were beaten. And that is another way of saying that we won.'

From the worm's eye view of a pilot officer in a fighter squadron at the time it all seemed quite simple. In September 1939 the Germans had obliterated Polish resistance by ruthlessly effective use of air power followed by swift-moving mechanized ground forces – the 'blitzkrieg' technique. They had just done the same in Western Europe, occupying Holland, Belgium and France in a matter of days, having first of all established total air superiority. That left England. You did not need to be a Clausewitz or Liddell Hart to guess what was coming next.

For those who want the statistics, they are all available from the official documents. For those who want opinions about the conduct of the Battle, on all sides, there are millions of words to choose from. But only those who were there, alive in Britain in that high summer of 1940, will ever really know what it felt like.

It was not just on the south coast, in sight almost of the enemy, that people felt conscious of the imminence of danger. In every far valley and village throughout the kingdom men and women of all sorts and condition prepared to defend themselves and their homes and their country.

I went home, after Dunkirk, for four day's leave in our little West Riding village of Cawthorne. There, in that patch of hilly, stone-wall country bordered on the west by the Pennines and ringed on other sides by the industrial and coal-mining belt of West and South Yorkshire, preparations to resist a German landing went forward on all sides. In every village, Cawthorne included, middle-aged and even quite elderly men formed units of the Home Guard, then know as Local Defence Volunteers. I found my sixty-three-year old father anxiously contemplating the respective merits of handing over his sporting guns to supplement the local unit's armoury of motley weapons, or of keeping them to use himself if necessary. He finally reached the prudent conclusion that the latter course was to be preferred. Signposts in the steep, narrow lanes were removed to confuse the Germans when and if they arrived. Obstacles were erected on all level places – and there are not many level places in that part of the world – to obstruct the landing of airborne troops.

I went over to nearby Wortley, to Lord and Lady Wharncliffe's house. I went really to see Diana and Barbara, the two of their four daughters who were closest to my age. No sooner had I got there than Archie Wharncliffe drove me out to help him with the placing of obstructions on possible landing sites on his estate. One such site which occupied his special attention was the golf course – his golf course, he called it, for it was laid out on his land, across the valley, on the far side of the park to the front of the house. Now he was determined to make quite sure that it would not provide landing space for Germans. Trees were being cut down to provide obstructions on the fairways. A notable feature of the course was its exceptionally hilly nature, I pointed out that this alone would make it almost as difficult for a pilot to land on as it undoubtedly already was for a golfer to golf on. But Archie would have none of that, nearby Sheffield, he insisted, with its steel works and other industries, would certainly be a priority objective. Sheffield was only five miles away, Rotherham only eight. He would do all in his power to ensure that the bloody Germans could not launch their attacks on these places from his golf course.

That was the spirit of those times, in the northern valleys no less than on the Channel coast. Most people expected that the Germans would attempt an invasion. And even though the general reaction to that expectation was one of more or less optimistic defiance, a sense of foreboding hung over the whole country.

My own family was faced by a pressing problem. My youngest sister, Charmian – then only nine years old – was at a school which was being moved lock, stock and barrel to Canada. Should she go with it? I was brought into the family deliberations. And I voted that she should go. I did so because I genuinely believed it extremely likely that the country would be invaded and become a battlefield. If it had simply been a question of bombs and all the other unpleasant dangers and difficulties which, at the very best, were so evidently to affect everyone in Britain, I would have said that she should stay. But no; the Germans would try to invade and might well succeed in doing so. I felt pretty sure about that, in the immediate aftermath of what I had seen over Dunkirk.

Back at Leconfield, however, we simply returned to the old routine of readiness and convoy patrols. But our enthusiasm was sharpened both by the awareness that the Germans were certain to attack Britain, in one way or another, within a few weeks and also by periodic brushes with enemy planes which harried the convoys and probed the east coast defences.

The squadron made two interceptions in June. Soon after we got back from Rochford, Jack Bell shot down a Heinkel III. And early in the morning of 29 June Roy Marples scored one of the very few night interceptions achieved by Spitfires, shooting down a Heinkel over Hornsea. This was a magnificent performance; but Marples, a young short-service commission officer, who was renowned for nocturnal operations of a different and more enjoyable kind, was found guilty by the rest of us of sabotaging our efforts to persuade the authorities that night patrols in Spitfires were a waste of time. We said that there must have been a woman in the German plane and that was the only reason why he saw it.

The Spitfire was, from the pilot's point of view, far from ideal for night flying under wartime blackout conditions. With the tail down in the landing or take-off position the long, broad engine cowling in front of the cockpit reduced forward vision so that you could only see out at an angle of about forty-five degrees on either side. The single line of 'glim-lamps', which was all we were allowed as a flare path, was masked in such a way that the lights could only be seen from a narrow angle on the line of approach. They did not illuminate the landing strip at all, being intended merely as a row of pinpoints on which to line up during the approach and keep straight after touchdown. Landing a Spitfire on a dark night guided only by those dim little lights was a hair-raising business for anyone. For inexperienced pilots it was potentially lethal.

On 7 July came the big moment when I at last scored hits on an enemy plane. Green section was scrambled and sent off at full speed out to sea, crossing the coast north of Spurn Point. George Moberley was leading; Sergeant Burnard and I followed him. There was a layer of broken cloud at about five thousand feet and we were ordered to keep below it. We thundered along at full throttle, bumping violently in the turbulent air just below the cloud. George gave the tally-ho and altered course sharply. Then I saw it – the pencil shape of a Dornier 17 twin-engined bomber just below the cloud, stalking a convoy.

As George went into the attack the Dornier pulled up into cloud. I climbed hard and in a few seconds burst through into the bright blue above. Almost immediately the Dornier also popped through close by and I was able to get in a short attack before it again disappeared into cloud. There followed an exciting chase as the German pilot tried frantically to elude us. But nowhere was the cloud solid, he was bound to come out into gaps and by good fortune we maintained contact with him, worrying at his heels like spaniels hunting in cover. He fought back gallantly – desperately would perhaps be a more appropriate word – and for a time his rear gunner returned our fire, though it was an unequal exchange, which must have been utterly terrifying for him. His tracer bullets streamed past and I received a hit on the outer part of my port wing. But the advantage was all in our favour. The rear gunner was silenced and the dying Dornier descended in its shroud of black smoke, to crash into the sea a few miles east of the convoy.

A second Dornier was sighted, scurrying away among the clouds. Sergeant Burnard and I managed to get in one attack each before our ammunition was all gone, but though we damaged it we did not see it crash.

Back at Leconfield I experienced for the first time the exhilaration of landing and taxi-ing in after a successful engagement with the enemy. Those who waited on the ground could always tell when a Spitfire's machine guns had been fired. Normally the eight gun-ports on the leading edge of the wing were covered by little patches of canvas. But when the guns were fired these patches were, of course, shot away, leaving the ports open, and the plane made a distinctive whistling noise on the glide. This clear signal that you had been in action could be made more pronounced by a bit of side-slipping, which, though sternly discouraged by the authorities, was hard to resist on such occasions. And so, when they recognized this signal of action, the ground crews, who identified themselves enthusiastically with the pilots whose planes they serviced, would run out in high excitement to hear the news. They regarded a victory for their plane as a victory for themselves – and justly so, for our reliance on their skills was absolute.

I felt twelve feet tall after that combat, which in retrospect certainly does not seem anything to be particularly proud of. At last I had broken my duck. I could only claim one-third of one enemy aircraft destroyed and one-half of another damaged – but that was better than nothing at all.

For the first time I was consumed by an insidious feeling which crept in on me many, many times in the months and years to follow. I heard the tempting tone of an inner voice which I was to hear again so often the next five years. It said: 'There, now. You have been in action several times and you have done some damage to the enemy. You are still alive and kicking. Even if you pulled out now, no one would ever be able to say you had not done your bit.'

It was the voice which expressed a sincere desire to stay alive, opposing a sincere desire to engage the enemy. It was muted and easy enough to muffle at that stage. But I was to learn how insistent it could become.

While we were still engaged with these comparatively low-tempo operations on the north-east coast, the pace was quickening for the squadrons in the south.

The key date in considering the story of the Battle of Britain is 16 July. On that day Hitler, after weeks of prevarication, finally made up his mind and set moving the train of events intended to culminate in the occupation of the British Isles by a foreign power for the first time in 874 years.
He issued his War Directive No.16. It said:

> As England, in spite of her hopeless military position, has so far shown herself unwilling to come to any compromise, I have decided to begin preparations for and, if necessary, to carry out the invasion of England.
>
> This operation is dictated by the necessity to eliminate Great Britain as a base from which the war against Germany can be fought. If necessary the island will be occupied . . . I therefore issue the following orders:
>
> 1. The landing operation must be a surprise crossing on a broad front extending approximately from Ramsgate to a point west of the Isle of Wight. . . . The preparations . . . must be concluded by the middle of August.
>
> 2. The following preparations must be undertaken to make a landing in England possible:
>
> (a) The English air force must be eliminated to such an extent that it will be incapable of putting up any substantial opposition to the invading troops. . . . .

There followed general instructions relating to the clearing of British minefields, the laying of German minefields and the disposition of heavy coastal batteries.

The directive was given the codename 'Operation Sea Lion'.

Look again at the wording of this historic document. First, the intention – subdue Great Britain, if necessary by invasion; second, the method of execution – eliminate the RAF.

Eliminate the RAF. That was the first priority and the task which had to be fulfilled before the bigger and final objective could be attained. And there is probably little doubt that Reichsmarschall Herman Goering, Commander-in-Chief of the Luftwaffe, assured his Führer that this could be quite quickly and easily achieved. The Luftwaffe, after all, had never known defeat. Over Spain, Poland, Holland, Belgium and France, German pilots and planes had operated almost at will, easily brushing aside all opposition from enemy air forces, laying whole cities to waste in a few hours of concentrated bombing, spreading terror and devastation along the crowded highways, where the military transport of defending ground troops became hopelessly and helplessly entangled and enmeshed with masses of civilian refugees.

Why should the all-conquering Luftwaffe now fear a setback? True, the Hurricanes they had met over France and the few Spitfires over Dunkirk had turned out to be deserving of respect. But what could a handful of such planes do against the vastly more experienced and numerous pilots and planes of the Luftwaffe? A glance at the order of battle as it existed at the time when War Directive No. 16 was issued is enough to show that Goering and his airmen had grounds for confidence.

Sir Hugh Dowding had ready for action twenty-two squadrons of Hurricanes and twenty squadrons of Spitfires – a total of about seven hundred and fifty planes, including reserves. That was the front line force and it had to defend the whole of the British Isles. A further eight squadrons of Hurricanes were forming or re-forming. And in addition there were a few squadrons of twin-engined Blenheims and rear-gun Defiants, both useless in the day fighter role.

Across the Channel, ranged from the Dutch Islands to Britanny, were two German air fleets under Kesselring and Sperrle. Between them, they had available about 2460 warplanes in the front line – 1200 twin-engined bombers, 280 Stuka dive-bombers and 980 fighters. A third air fleet was positioned in Norway, ready to attack across the North Sea against targets anywhere along the eastern seaboard of the British Isles. This third German force consisted of about a hundred and ninety planes, a mixed assortment of bombers and long-range Messerschmitt 110 fighters.

Thus the overall numerical superiority of available German planes was a little more than four to one. It may have seemed a nice degree of superiority to the Germans as they prepared for the onslaught, particularly when it is remembered that less than two-thirds of Dowding's planes were within range of the south coast battle area, while every one of the German fighters – except those in Norway – was based within range of London. But it was not in fact a degree of superiority which would have commended itself to a more cautious and less vainglorious commander embarking on an exercise of such magnitude. In particular it may be seen that, judged by experience gained later in the war, the proportion of fighters to bombers and the degree of overall fighter superiority were both unsatisfactory.

What about the quality of the German planes? Undoubtedly – and it was no less obvious to all of us at the time then than it is with the benefit of historical hindsight – the Luftwaffe in 1940 possessed the finest and most powerful fleet of warplanes existing in the world at that or at any previous time. Nevertheless, in the squadrons we had no great fear of the German bombers. Every fighter pilot's dream was the opportunity to find a formation of unescorted Stukas, the slow, lightly-armed, single-engined Ju 87 dive-bombers which had spread so much fear and devastation over the continent. The Stuka was an extremely effective piece of aerial artillery which had been used with devastating effect in Poland and Western Europe. But it had to rely absolutely on an effective fighter screen for defence against Hurricanes and Spitfires.

The twin-engined bombers – mainly Heinkel IIIs, Junker 88s and Dornier 17s – were fast by the standard of the day. But they were indifferently armed. They could rarely bring more than one gun at a time to bear on an attacking fighter. Their designers had intended that they should rely more on speed than on fire power for their defence; and this theory would have worked very well against the Gauntlets, Gladiators and Furies with which the RAF fighter squadrons had, in the main, been equipped until 1939. But they were not fast enough to evade our Spitfires or even the much slower Hurricanes and we knew we could deal with them effectively if we could just get at them.

Our feelings towards the Messerschmitt 109, the principal single-seater fighter in the German air force were very different. The Me109 was a most effective fighter indeed and it accounted for most of the losses suffered by Fighter Command during the Battle. It was as fast as the Spitfire, considerably faster than the Hurricane and could out-dive and out-climb either. Its armament was formidable, as many of us were to discover. There was a cannon, firing explosive shells through the propeller hub. And there were four or six heavy machine-guns, of which two were mounted above the engine cowling.

Half a dozen shells from these guns could do great damage – much more than the equivalent number of hits from our own Browning machine-guns. On the other hand, the firing rate of the Brownings was much higher, which gave us a better chance of scoring with a short burst of fire.

In one vital respect the Me109 was at a disadvantage against the British airplanes. It could be out-turned both by the Spitfire and the Hurricane. This was a serious handicap to the Luftwaffe pilots allotted the duty of providing close escort for the bombers. Their freedom of action was curtailed. They could not pursue the tactic, best suited to their planes, of a high-speed attack followed by dive or zoom. They had to try to stick around and fight it out; and that involved the matching of turning circles. They never found a way round that problem and their difficulties were made all the greater when Goering, infuriated by the losses inflicted on his bombers, ordered the fighter squadrons to cling ever closer to the bombers they were escorting.

Some of the Me 109s were painted a glaring, garish yellow from propeller hub to cock-pit. Although this made them easier to see it also made them appear more frightening. Rightly or wrongly we assumed that the yellow-nose airplanes belonged to some particularly crack formation.

Another alarming feature of the 109s was that they used tracer ammunition. This, in fact, probably rebounded to our advantage, because it could warn the unwary pilot that he was being shot at and might give him a split-second chance to take evasive action. But I always thought, personally, that bullets you could see were worse than bullets you could not see. And the spectacle of yellow noses hosing out orange-coloured tracer was particularly unsettling for the nerves.

There is no doubt, however, that Goering and his commanders overrated the effectiveness of their own fighters in relation to our own. In fact the Messerschmitt 109 and the Spitfire were extraordinarily evenly matched. Their duel for supremacy lasted throughout the war, as each plane was constantly improved and given increased power and performance. At times the Germans, by rushing out a new version before our own next improvement was ready, would get one jump ahead. At other times the advantage would be to the RAF. But on balance the Spitfire was, I believe, slightly the better basic aircraft. And so it was in 1940. In particular, such advantages as it enjoyed over the Me 109 at the time were enhanced by the circumstances of the battle.

The Hurricane was in a somewhat different category. For fighter-versus-fighter work it was already obsolescent in 1940. It did not have the speed or rate of climb to compete with the Me 109 on level terms. But here again the circumstances of the battle – the fact that many of the Messerschmitt squadrons were under orders to stick closely to the slow bomber formations – rebounded to our advantage. Never again, after 1940, were Hurricanes used with success in large scale operations against German fighters.

But in the Battle of Britain they played a successful and indeed a glorious role.

CHAPTER FIVE

Battle of Britain

Historians have never been able to agree about the date when the Battle of Britain started. The question is academic; but certainly mid-July marked the beginning of Fighter Command's ordeal and, in particular, the ordeal of No.11 Group and its New Zealand commander, Air Vice Marshal Keith Park.

Fighter Command at that time was divided into four groups, each with its own geographical area of responsibility. Park had the hottest seat; his command was bounded by a line which ran south-west from the coast of East Anglia at the Norfolk/Suffolk border, passing to the north of London and veering south to the English Channel just west of the Isle of Wight. Thus the capital city, the home counties and several important sea ports fell within his area, and because his airfields were closest to the enemy bases he had the shortest time in which to make his decisions.

To the west of 11 Group, guarding the western part of the Channel, was the newly formed 10 Group, under Air Vice Marshal Sir Quentin Brand. The squadrons of 10 Group were frequently called on to help defend Portsmouth and Southampton, which were close to the boundary dividing the two groups.

To the north of London, stretching as far as our own sector in South Yorkshire, was Air Vice Marshal Trafford Leigh-Mallory's 12 Group. His southernmost sector stations were Duxford and Coltishall, near Cambridge and Norwich respectively. To the north again, reaching all the way from Flamborough Head to the Orkneys, was Air Vice Marshal R.E. Saul's 13 Group.

The strain imposed on 11 Group was tremendous. It was Dowding's inflexible policy to allow Park only about forty per cent of the total force available — between twenty and twenty-five squadrons. And although of course the squadrons were rotated between the groups, the members of Park's own headquarters organization and staff remained at the forefront throughout those two and a half months of crisis, working from before dawn until dusk, seven days a week.

In mid-July the Germans, still unprepared to begin a major assault against

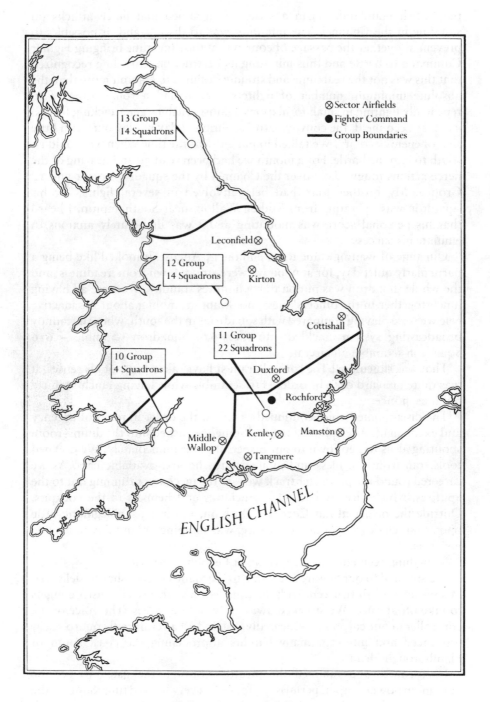

Fighter Command, July 1940

the British mainland, began a series of sustained and fierce attacks on shipping in the Channel. The intention was to disrupt and, if possible, to prevent altogether the passage of convoys, at the same time bringing Fighter Command to battle and thus initiating its destruction. Dowding recognized that this was not the real thing and stolidly declined to commit more than the absolute minimum number of fighters. Those had a rough time of it, repeatedly fighting in small formations against much larger attacking forces.

Up at Leconfield, the convoy patrols continued day in, day out, with little sign of enemy activity. We talked impatiently of the time when we would go south to join the battle. For a month we had been reading and hearing of the fierce actions fought daily over the Channel by the squadrons of 11 and 10 Groups. My brother John had been involved in several fights, for his squadron was operating from Middle Wallop, near Southampton. I heard that his personal score was mounting and I was desperately anxious to emulate his success.

Our time of waiting came to an end on 15 August. It looked like being a particularly quiet day, for at noon we were stood down from 'readiness' and the whole squadron was put on thirty minutes standby. We were all having lunch together in the officers' mess, no doubt grumbling about the inactive role we were playing compared with squadrons in the south, when the tannoy broadcasting system crackled into life: '616 Squadron scramble – 616 Squadron scramble all aircraft.'

That was ridiculous! The controller must have taken leave of his senses. It was quite unheard of to be ordered to scramble while having lunch at thirty minutes' notice.

The disembodied voice continued to repeat the order in tones of urgency and excitement. A telephone rang and someone rushed into the dining-room shouting at us to get down to our dispersal point immediately. We downed tools, ran from the mess and jumped into the first available cars. As we careered round the perimeter track we saw the mechanics running out to the Spitfires, which already had our parachutes and helmets in the cockpits. Outside the dispersal hut Corporal Durham, the usually phlegmatic little operations clerk from South Yorkshire, was jumping up and down, waving his arms.

Something seemed to be happening at Leconfield at last.

We sprinted to our Spitfires, reacting to Durham's urgent shouts, delivered in a broad Yorkshire accent with the usual epithets thrown in, instructing us to take off at once. We streaked away in twos and threes. The voice of the controller in our earphones repeatedly ordered all aircraft to fly out to sea at top speed and intercept many bandits approaching the coast south of Flamborough Head.

I set course and rammed the throttle 'through the gate', to get the maximum power output, permissible for only a very limited time. Some of the others were ahead of me, some behind. We did not bother to wait for each other or try to form up into flights and sections. We raced individually across the coast and out to sea. About fifteen miles east of Bridlington I saw them, to the left front, and slightly below – the thin, pencil shapes of German twin-engined bombers, flying in a loose, straggling, scattered formation toward the coast.

I switched on my reflector sight, setting the range for two hundred and fifty yards, turned the gun button to the 'fire' position. Wheeling down in a diving turn, I curved towards the nearest bomber, judging my rate of turn and dive to bring me in astern. A light winked from the rear-gunner's position and tracer bullets hosed lazily past. When I opened up with my eight Brownings the return fire stopped. The bomber turned and lost height. First a gush of black smoke, then a steady stream poured back from its engine cowlings and it fell steeply towards the calm summer sea.

Turning to look for a second target I saw other Spitfires fastening on to the German planes on all sides. Beneath me a damaged bomber turned back out to sea and I decided to go in and finish it off. It was a foolish decision, made in the heat of the moment, for I should have looked for an undamaged plane still making for the coast. By the time I had caught up with it and knocked it down I was several miles further out to sea. The sky was empty and I judged that my ammunition was nearly exhausted.

Hot and elated I set course back to base. I touched down only twenty-five minutes after taking off. The rest of the squadron aircraft were straggling in, singly or in pairs. Everyone seemed to have fired his guns.

The squadron claimed eight Junkers destroyed – of which I personally claimed one-and-a–half – four probably destroyed and two damaged. Not one of our planes was even damaged. Only a small number of the enemy aircraft penetrated inland. They made an abortive attack on the bomber base at Driffield, a few miles north of Leconfield. One Junkers dropped its load on a bomb-dump, causing a major explosion. But only minor damage was caused to the airfield itself. And the Whitley bombers based there were able to carry out their own raid against Germany that night as scheduled.

We did not know it at the time, but we had in fact played a small part in warding off a blow which Goering intended as the killer punch of the RAF. In planning the operations of 13 August, Goering named the day 'Adlertag' – Eagle Day. His confident assumption had been that on that day his German Eagles would swoop on Dowding's already ruffled sparrows and deal them such a blow that all resistance would break within a week. The Reichs-marschall's confidence was based on a misinterpretation of the effects of his decision, implemented four days earlier, to move the weight of his attacks inland from the Channel. Main targets were Fighter Command airfields and radar stations. German documents show that Goering was led by over-optimistic intelligence reports to believe that, as a result of raids carried out on 12 and 13 August, radar defences in the south had been drastically impaired, that eight major airfields had been put out of action and that the Luftwaffe had gone a fair way towards destroying the 11 Group fighter forces. He was confident that reinforcements would have been sent south from 12 and 13 Groups, leaving the fighter defences of northern Britain much depleted.

Accordingly, on 14 August, he gave his bomber squadrons a day off to prepare for the knock-out blow. And the next day he sent them out again in maximum strength – not only across the Channel but across the North Sea also, expecting to find the north-east and Midland areas denuded of fighters. His unfortunate aircrews got a nasty surprise, as the Spitfires and Hurricanes in Yorkshire, Durham and Northumberland rose up against them. And in the south things were no easier for the raiders than they had been on the preceding days.

One thousand eight hundred German aircraft flew against Britain on 15 August. Seventy-six of them were destroyed, for the loss of thirty-four British fighters.

As a killer-punch it was something of a failure – the Eagle was losing more feathers than its prey.

The explosions were so unexpected, so shattering, their effect on my Spitfire so devastating, that I thought I had been hit by our own heavy ack-ack.

White smoke filled the cockpit, thick and hot, and I could see neither the sky above nor the Channel coast twelve thousand feet below. Centrifugal force pressed me against the side of the cockpit and I knew my aircraft was spinning. Panic and terror consumed me and I thought, 'Christ, this is the end.' Then I thought, 'Get out, you bloody fool; open the hood and get out.'

With both hands I tugged the handle where the hood locked onto the top of the windscreen. It moved back an inch, then jammed. Smoke poured out through the gap and I could see again. I could see the earth and the sea and the sky spinning round in tumbled confusion as I cursed and blasphemed and pulled with all my strength to open the imprisoning hood.

If I could not get out I had at all costs to stop the spin. I pushed the stick hard forward, kicked on full rudder, opened the throttle. Nothing happened. The earth went spinning on, came spinning up to meet me. Grabbing the hood toggle again, I pulled with all my might, pulled for my life, pulled, at last, with success. I stood up on the seat and pushed the top half of my body out of the cockpit. Pressed hard against the fuselage, half in, half out, I struggled in a nightmare of fear and confusion to drop clear, but could not do so.

I managed to get back into the cockpit, aware now that the ground was very close. Try again; try the other side. Up, over – and out. I slithered along the fuselage and felt myself falling free. Seconds after my parachute opened I saw the Spitfire hit and explode in a field below. A flock of sheep scattered outwards from the cloud of dust and smoke and flame.

For a few moments there was silence and peace. Then the ground swung up fast and I remembered to bend my knees and roll over and bang the quick-release catch of my parachute harness. I lay under a hedge by the side of a wood. Two or three hundred yards away my Spitfire burned. My left leg was sticky with blood and my left shoulder, badly dislocated, hurt abominably.

A farmer with an old-fashioned hammer-gun stood over me and I thought his attitude none too friendly. Probably he did not much like having aeroplanes making holes in his fields and frightening his sheep.

The next few minutes are a blur in my memory. A number of people in khaki appeared on the scene and there was a certain amount of to-ing and fro-ing before I was driven away in an Army ambulance to the Kent and Canterbury Hospital. Later I learned that my tattered and bewildered person had been subjected to something in the nature of a tug-of-war between Captain Max Wilson, Medical Officer of the 64 Field Regiment, Royal Artillery, and Mr George Henbury, local builder, Home Guard Commander and all-round bigwig. The latter wanted to take me off to have a drink or two with his cronies; the former insisted that I was in no condition to be thus paraded and entertained and must be taken straight to hospital. Fortunately for me, Captain Wilson got his way and George Henbury lost his prize.

Next morning I learned the humiliating truth: that I had been shot down by a Messerschmitt 109 which I had not even seen.

That unpleasant and nearly fatal event had followed soon after our eagerly awaited move from Leconfield to the south, on 19 August, four days after our successful engagement on Eagle Day. We had been sent down to relieve a battle-weary squadron at Kenley, a sector station in Surrey, on the southern outskirts of London.

Joy and jubilation marked our last hours at Leconfield. We did not know that they were also to be the last hours in which the squadron would exist in its old carefree form. Most of the original auxiliary pilots were still there – Teddy St Aubyn, leading the festivities in the mess at lunch time; George Moberley, making arrangements for the care of his much-loved dinghy, *April*; Jack Bell, Ken Holden, Dick Hellyer, 'Buck' Casson, John Brewster – we had come through together from the earliest days of 616 squadron and it never occurred to us that we should not continue together indefinitely. And so we drank a little more than usual at lunch time and went down to the airfield afterwards in a hilarious mood, eager to take off for Kenley and glory.

When we landed at our new airfield our spirits were quickly subdued. Kenley had been heavily and successfully blitzed the day before. Much of the station lay in ruins. Wreckage of aircraft and motor transport spattered the periphery of the field. Newly-filled craters dotted the landing ground.

The atmosphere in the officers' mess was taut and heavily overlaid with weariness. Both the station operations staff and the pilots of 615 squadron, led by Squadron Leader Walter Churchill, showed signs of strain in their faces and behaviour. The fierce rage of the station commander when a ferry pilot overshot the runway while landing a precious replacement Spitfire was frightening to behold.

Our arrival coincided with a couple of day's break in the battle. We sat at readiness throughout the long daylight hours from early morning to late evening. We scrambled once or twice, but made no contact with the enemy. Soon after tea on 22 August we were put on thirty minutes' notice and told that a little later we would be released until dawn next day. Back in the officers' mess we changed into our best uniforms and planned a night out in the West End of London. But no sooner were we ready to go than the tannoy called us back to readiness. Cursing our luck, we drove round the perimeter track to dispersal, lugged our parachutes out to our planes, put on Mae West life jackets over our tidy tunics and awaited developments.

It turned out that the reason for our recall was a surprise visit to the station by the Prime Minister, Winston Churchill. That, at least, made the sudden and enforced change in our evening plans seem worthwhile. A small procession of cars drove round the taxi track and stopped outside our hut, where we lined up to meet the great man, the lion who roared for England.

Just a few seconds after I had shaken hands with Churchill, Corporal Durham appeared on stage to produce one of his more spectacular performances. Racing out of the hut, he shouted to us to scramble. I remember thinking as we sprinted out to our planes and streamed down the runway in the fastest take-off of the squadron's history that the whole performance had obviously been laid on for the Prime Minister's benefit. Unfortunately it was in that unexpectant frame of mind that I flew along, 'tail-end Charlie' in our formation of twelve, when, a few minutes later, we levelled out twelve thousand feet above Dover. I was still wondering whether there might yet be time to get into the fleshpots of London when the cannon shells from that unseen Messerschmitt ripped home and sent me spinning on my first face-to-face meeting with death.

From that evening until mid-September I was out of the cockpit. Thus I missed the desperate, climactic stage of the Battle of Britain which took place in the last week of August and the early days of September. Like millions of other people on the ground in south-east England I watched the furious conflict in the summer sky. We had a grandstand view from the windows and terraces of the hospital just outside Canterbury. Every day more wounded pilots came in and they brought news from the squadrons. And every day, as I sat uselessly on the ground, I heard of friends who had died and others who were wounded.

On 25 August George Moberley came to see me. He had been in action that morning, when he had shot down a Messerschmitt 109. In the same action Sergeant Ridley had destroyed a Dornier 17, Jack Bell had claimed a Messerschmitt as probably destroyed and three more Messerschmitts and one Dornier had been claimed as damaged. George told me how Sergeants Westmoreland and Waring, both 'B' Flight pilots, had been shot down that day. Westmoreland was known to be dead; Waring was missing. Then he talked to me about personal affairs, about his family and his property. He told me that he wanted me to have his personal belongings if he were killed. I had a strong feeling that he had a premonition that he would be.

If indeed he had such a premonition, it was fulfilled next day, when, in a dogfight south of Dover, George was shot down and killed. He baled out at a very low altitude – so low that his parachute could not open properly. He hit the sea close to shore and his body was recovered.

In the same action Sergeant Ridley was killed and Teddy St Aubyn was shot down and very badly burned. For days his life hung in the balance and he was in hospital for several months. Later, despite his terrible wounds and his age – which was far above that normally accepted as the maximum for fighter pilots – he managed to get back to operations as a pilot in the army co-operation squadron equipped with Mustang fighters. He was killed in action in the early summer of 1944, when, leading a section of planes back across the Channel at wave-top level after a sortie over France, he misjudged his height and hit the water.

Roy Marples, Bill Walker and Sergeant Copeland were also shot down and seriously wounded on that black day for 616 Squadron. Four days later Jack Bell was killed; he was badly shot up in combat and evidently he himself was grievously wounded, for he tried to force-land at West Malling but crashed and burned out on the airfield. The indications were that he had probably passed out in the final stages of his landing.

In eight days, 616 Squadron lost five pilots killed or missing and five others wounded and in hospital. About fifty per cent of the pilots who had flown down from Leconfield had gone. But still the remnants fought on, greatly inspired by the courage and leadership of Denis Gillam, who now showed the fiery and utterly fearless character which was to make him one of the most distinguished RAF officers of the war. But on 2 September he too was shot down and injured. It was the final, knock-out blow. Without him the squadron was incapable of continuing the fight. And so, exactly fourteen days after our carefree and confident departure from Leconfield, 616 Squadron – what was left of it – was taken out of the line to re-form at Coltishall, near Norwich. To set against its losses, the squadron claimed sixteen enemy aircraft definitely destroyed, six probably destroyed, fifteen damaged. Of these Gillam had personally destroyed seven and damaged three.

While the squadron was at Coltishall Marcus Robinson was replaced as Commanding Officer by Squadron Leader Billy Burton, a Regular officer with impeccable credentials – not only a Cranwell man, but a Cranwell Sword-of-Honour man. At the same time Dick Hellyer left to take up new duties as a flying instructor.

CHAPTER SIX

The Bader Factor

The story of 616 Squadron from 19 August to 4 September is by no means untypical of other fighter squadrons at that time. Of course, there were many squadrons which did much, much better. Others did worse.

The important thing is that at the end of the day most squadrons could claim that they had given a little better than they had got. If they had lost ten airplanes and pilots they had destroyed fifteen or twenty. That was the ratio which mattered. Scores of unknown young pilots who, in the last moments of their lives, never knew what hit them, had succeeded first in scoring a victory or two. And that, in the final analysis, was why the Battle of Britain was won.

But at the time when 616 Squadron was pulled out of the line, while I was still convalescing from my minor wounds, the battle was far from won. It was, indeed, the darkest hour, the period of greatest strain for the RAF and of greatest peril for the country. During that period the Luftwaffe pursued the line of action most likely to win the day. Goering had issued a directive to his chief lieutenants, Kesselring and Sperrle. 'Until further orders,' it read, 'operations are to be conducted exclusively against the enemy air force ... for the moment other targets should be ignored.'

By the end of the first week in September the policy was beginning to pay off. Day after sunlit day an average of one thousand German airplanes came over. Dawn after chilly dawn the weary British pilots assembled at their dispersal points and waited quietly for the telephone call which would send some of them to death, even before breakfast. Night after weary night the reckoning was made and though the advantage was constantly in Britain's favour and though no doubt the German pilots were almost as bone-weary as our own and the Luftwaffe's morale must have been severely affected by the daily loss of dozens of crews and the grisly spectacle of many more aircraft returning riddled by bullets and soaked in blood – yet the steamroller technique was beginning to tell against Dowding and England.

The supply of pilots began to dry up. Some were shot down two or three times but, escaping injury, returned to the battle. Others were killed before they had fired a shot. Most survived a few days before falling in the fury of the fight, either to their death or to a period of convalescence from their wounds.

Dowding could not rotate his squadrons fast enough to keep pace with the losses. Squadrons in the south became depleted before others, taken out of the line to re-form, could build up their strength again. Dowding had to take experienced pilots from the squadrons which were resting and re-forming, in order to plug the gaps in other squadrons, which should really have been taken out of the line. It was a policy of desperation and it could not last for long. In the darkness of that crisis it may well have seemed to our fifty-eight-year old commander that it was a problem without a solution.

It might have been so, but for the intervention of Hitler himself, who now had one of those flashes of intuition which, from time to time, brought such dire consequences to his country. At the moment when the battle was in the balance, when the weight of Goering's strategy was coming close to success, when Fighter Command was near to breaking point – at that precise moment of crisis something else broke. It was Hitler's patience. The Führer spoke and Goering obeyed. The point and purpose of the German attack was diverted from the destruction of the RAF to the cowing and subjugation of London.

It was the turning point. London burned; but Britain was saved.

The first big attack on London came on 7 September. And it is certain that the pilots of Fighter Command felt no sense of relief as the assault on the capital built up to its mighty climax on 15 September. On the morning of that day the Prime Minister, Winston Churchill, visited the 11 Group Operations room. As he sat and watched, the plots showing enemy formations started building up on the table below him. To Air Vice Marshal Park it was soon evident that a raid of extraordinary strength was assembling. For several minutes the plots stayed over France, constantly increasing. The German planes gathered in their hundreds before moving north across the Channel and thence across Kent and East Sussex towards London.

Every available Spitfire and Hurricane in 11 Group was scrambled. And Park called in the five 12 Group squadrons which had been assembled at Duxford under Squadron Leader Douglas Bader. At the height of the battle which developed – the fiercest and most concentrated of the whole campaign – it is reported that Churchill turned to Park:

'How many fighters have you left?' he asked.

'None, Sir,' Park replied.

And that was the margin.

Over the years I have been constantly aware, when, as happens from time to time, people – well-meaning, no doubt, but not very well-informed – refer to me as though I had played some distinguished part in the Battle of Britain, that while all these momentous events were taking place I was on the ground, first in hospital and then recuperating at home in Yorkshire with my mother and father. I left hospital on 28 August and arrived at Cawthorne the next day.

I have little detailed recollection, and no written record, of my sick leave. I remember that there was an army unit camping in the park at nearby Noblethorpe, the home of Mrs Fullerton (a well-loved honorary aunt) and that I had some convivial times in their officers' mess tent. And I remember a morning's shooting at Ardsley, with Gerry Micklethwait and one of his wife's nephews, whom I sought to impress, during intervals in our chivvying of the few partridges present in those fields on the south-easterly extremities of Barnsley, by passing off my recent misadventure in a suitably dashing and Biggles-like way.

I was at Cawthorne for a fortnight and then had to report to Kenley for a medical examination, when I was passed fit to return to duty. And so, on 13 September, I drove north in George Moberley's old 8 h.p. Ford and rejoined 616 Squadron at Kirton-in-Lindsay, where it had been sent to reform.

Kirton was a bleak and windswept airfield on the edge of an escarpment eleven miles north of Lincoln. Bleak, too, were my feelings when I walked into the pilots' hut and looked around. In 'A' Flight Ken Holden was still there, and John Brewster. In 'B' Flight there was 'Buck' Casson. One or two sergeant pilots remained from the Leconfield days. All the other faces were those of strangers.

Billy Burton, the new CO, was not a man to tolerate a situation or an atmosphere where there could be any brooding over the past. He sent for me and asked me straight out whether I felt in every way fit to go back to flying. It was a critical moment. The memory of my terrifying experience on 22 August was overpowering. And it was reinforced by the absence of so many of my old friends and the knowledge of what had happened to them. If I now intimated privately to Burton that a longer break would be welcome it would probably be arranged (I later learned that Leigh-Mallory was looking for a new PA and that my name had been mentioned) and I would incur no criticism. The inner conflict between a strong desire to stay alive rather than to engage the enemy was very much to the fore at that moment. But I heard myself saying that, yes, I did indeed feel in every way fit and ready, whereupon Burton very sensibly ordered me into the air without further delay or discussion.

As soon as I had taken off all other feelings were driven out by the sheer joy of flight, which flooded back in. I looped and rolled and chased the clouds in delight. When I landed I was able to tell Burton in good faith that I really did feel fine and ready for anything.

Burton had been given two flight commanders to help him reshape the squadron. The senior of the two, who now became second-in-command of the squadron was Jerry Jones, a tough terrier-like professional who had been a Sergeant Pilot before the war. He commanded 'A' Flight. The new commander of 'B' Flight, to which I had returned, was Colin Macfie, an Auxiliary Air Force officer from 610 (Liverpool) Squadron. He was very young – still under twenty-one – exceptionally good-looking and reserved to the point of appearing withdrawn, but he was a pilot of rare skill. Aided by these two men, Burton set about getting the squadron operational again at top speed. Every day, several times a day, while the battle still raged over London, we flew hard, taking up the new pilots for intensive operational training.

On 18 September five days after my return, Burton told the station commander that we were ready and the operations staff at 12 Group acted on this information without delay. We were ordered to take off at dawn next morning to fly south to Fowlmere, a satellite field in the Duxford sector south of Cambridge, for operations with the 12 Group Wing. We would return to Kirton in the evening.

I viewed the prospect of combat with real inner fear. The memory of what had happened last time crowded back in on me. The juvenile desire for glory which had been uppermost in my mind before we went to Kenley had been driven out altogether by the fear of death and the personal knowledge of the unpleasant form in which it was likely to come. As we flew south through the sharp September early morning air I prayed that there would be no action for 616 Squadron that day.

The 12 Group Wing was a formation of five squadrons assembled by Leigh-Mallory to go into battle as one cohesive unit. It was led by Douglas Bader, the already legendary character who had talked his way back into the Service – and onto operational flying – at the outbreak of war, in spite of the fact that he had lost both legs when he had crashed in a Bulldog fighter eight years before. He was commanding 242 Canadian Squadron, which was equipped with Hurricanes and was flying into Duxford every day from Coltishall. Two Polish squadrons – 302 and 310 – were based at Duxford and they, too, had Hurricanes. Resident at Fowlmere was the famous 19 Squadron, equipped with Spitfires – the same unit by which I had been so impressed as they stood at readiness when we landed at Duxford to refuel almost exactly a year before, on our way back to Doncaster from Manston.

After we had landed at Fowlmere and seen our planes refuelled and made ready for action we went into a hut on the edge of the field for tea and sandwiches. There we met the pilots of 19 Squadron and settled down to wait for developments.

Quite soon the door was flung open and an extraordinary figure in squadron leader's uniform stomped vigorously into the hut. Instantly the subdued and somewhat queasy atmosphere was dispelled, driven away by the hard, robust character of Douglas Bader as he called noisily for Billy Burton, an old friend, and then greeted each one of us individually. Much has been written about Bader. I shall write much more, for we were to be closely associated during the spring and summer months of 1941. My personal debt to him is incalculable. He showed me quite clearly by his example the way in which a man should behave in time of war and his spirit buoyed me up through many dark days long after he himself became a prisoner of war.

With Bader on this occasion of our first meeting was the Duxford sector commander, Wing Commander 'Woodie' Woodhall. Then and later, in 1941, he and Bader combined together with remarkable effect – Bader as the formation leader, always, always looking for a fight; Woodhall as the controller with a cool, unruffled ability to size up the situation presented to him on the big board in his operations room and then to give directions to his pilots in a voice which invariably remained calm and sonorous, even at times of greatest stress.

I felt better, much better, after their visit. I was almost anxious to get airborne on patrol behind this fabulous character. But in fact nothing happened that day. The five squadrons sat on the ground inactive and late in the evening we flew home to Kirton. From then until the end of September we went down to Fowlmere most days, once or twice spending the night there in bunks in the dispersal hut. Almost every day we took off and patrolled the London area in our strong formation of sixty planes, the three Hurricane squadrons in a wedge below, the Spitfires above and to one side.

On the first of these sorties, after the five squadrons had assembled in proper order and were climbing away to the south and the butterflies in my tummy had begun to work overtime, Bader's voice rang out on the radio, calling Woodhouse in the control room. To my amazement the purpose of his call turned out to be the arrangement of a game of squash. Bader explained that he had intended to ask a particular person to play with him but had forgotten to do so before taking off. He asked Woodhall to do it for him. The conversation had a decidedly calming effect on my nerves and the butterflies were somewhat subdued. It was extraordinary enough that a man with two tin legs should have been thinking about squash, in any circumstances. That he should be doing so while leading three squadrons of Hurricanes and two of Spitfires into battle against the Luftwaffe was even more extraordinary. Here, quite clearly, was a man made in the mould of Francis Drake – a man to be followed, a man who would win.

I was subsequently to learn, at very close quarters, how true and accurate that first judgement had been. But I had to wait until the spring and summer of 1941 for that, because the Duxford wing, though often on patrol, was engaged in only one major action during that second half of September, at the tail-end of the Battle of Britain, when 616 Squadron was part of it. That action, the last big throw made by the Germans, took place on 27 September. Goering once again lashed out in daylight with all his strength. It is hard to see what drove him to do so, for the proposed invasion had by then been postponed until the following spring.

The targets on 27 September were London and the Bristol Aircraft factory at Filton. The two operations were launched simultaneously in mid-morning. About three hundred German planes were sent against London, more than two hundred of them fighters. These Messerschmitts appeared first with a small force of bombers, which were engaged before reaching the capital, dropping their bombs around the southern outskirts of the city. The fighter swarms meanwhile moved on to patrol over London while waiting for bigger formations of heavy bombers to arrive on the scene. The German plan, presumably, was to get this second formation through while the defending fighters were refuelling and re-arming. If so, the plan was a failure.

Squadrons of 11 Group intercepted the bombers over the coast and set about them to such good effect that they broke up in confusion. None reached further than the middle of Kent. Meanwhile we clashed with the waiting Messerschmitts over London.

Bader brought us in high, still flying in a solid wedge. We came together with the Messerschmitts in a monstrous explosion of planes and there developed immediately a dogfight of exceptional size and fury. I never again saw so many fighters engaged so closely together as in the first moments of that encounter. The effect for perhaps thirty seconds was one of complete chaos. It seemed impossible that one could avoid collision. The air was full of smoke and tracer as Spitfires, Hurricanes and 109s whirled around in desperate efforts to shoot each other and to avoid being shot. I turned and turned, sweating with exertion and excitement and half-sick with the fear of feeling again the thud of shells banging into my Spitfire. I took several snap shots, straightening out momentarily to take aim when an opportunity presented itself, pressing the button for a quick burst and flinging back into a steep turn before an enemy plane could pick on me. Then came the still unfamiliar miracle. The sky was empty. The tumult was over as abruptly as it had started. I put the nose down and headed north for Duxford.

I made no claim when I landed, beyond the assertion that if the German pilots were anything like me I had certainly frightened one or two. Only Ken Holden was certain of having scored a kill, for his victim had blown up before his eyes. Other pilots could do no more than claim that they had inflicted damage.

'Smithie', one of the Regular officers who had joined us at Doncaster on the outbreak of war, was killed. This final blow was most sorely felt.

The fact that this should have been the only major action in which the 12 Group Wing was engaged during the eight days we flew with it – a period when 11 Group squadron were in action fairly frequently – is relevant to a controversy which has given rise to much disputation between fighter pilots as well as historians of the Battle: was the 'Big Wing' concept a useful or a wasteful way of using five precious squadrons?

The bones of that controversy have been picked over all too often, to no one's benefit, and I shall leave them in peace. However, it is clear from my letters written at the time that, from the lowly position I then occupied, I perceived Bader and his 'Big Wing' as being almost a battle-winning factor. How I can seriously have reached such a conclusion when, at the same time, I was reporting that the Wing was seldom engaged with the enemy is difficult to understand; but my line seems to have been that its very presence over London was enough to keep the Germans away. Two letters written at our dispersal point on 25 September, one to my mother, one to John, express that view and their naïvety brings a blush to my cheek.

Our fight on 28 September was our last over London that year. The daily excursions to Fowlmere came to an end. And we settled down for the autumn and winter at Kirton.

CHAPTER SEVEN

Desolation

There has survived, from that long-ago winter at Kirton, a battered foolscap-size folder issued by the stationery office. Its brown, cardboard cover bears the printed words 'S.O. Duplicate Manifold Book Foolscap Folio' and, above them, the handwritten words 'P/O H.S.L. Dundas. Diary, Private.' On the bottom of the cover someone else has written 'Line Book, Master Copy'.

The first entry is dated 10 October 1940. Between then and January 1941 I wrote in the book quite regularly, on average at least two or three times a week. Then, like most young diarists, I lapsed. There are some quite long entries written in the spring of 1941, after we had moved south again. But it was only at Kirton, in the dying days of 1940, that there was any continuity. It is a revealing document and I was amazed, as I read it nearly half a century later, by the extent to which the flavour of the writing demonstrates my immaturity at that time. I was also amazed that I should have been so unashamedly introspective. It certainly is not a document in which it is possible for me to feel any pride – rather the reverse; embarrassment is a more appropriate word. But I quote it to tell the story of that winter, because it is, after all, the authentic record, unfiltered by time and unadulterated by selective memories.

To set the scene I should explain that, after the daily expeditions to join the 12 Group Wing came to an end, 616 was turned temporarily into a 'C' Class squadron. That was the role given, at that time, to most squadrons in 13 Group and also those, like ourselves, in the northern sectors of 12 Group. It was an expedient made necessary by the grievous wastage of pilots in 11 and 10 Groups. New pilots emerging from operational training units were posted to 'C' Class squadrons in each of which was maintained a hard core of more or less experienced men who could put the newcomers through an intensive course of battle training before they were sent to fill gaps in squadrons based in the south, where the Germans kept up a certain amount of activity, mainly in the form of fighter sweeps, throughout October and November.

The diary reveals a curious reaction to our squadron's role. It also records a number of meetings with my brother John, whom I saw more regularly during those few weeks than at any other time since before the war. The fact that his squadron, based at Middle Wallop, in Hampshire, was quite regularly in action possibly contributed to my own dismay at being confined to a 'C' squadron. The diary also described my reaction to the terrible news of John's death in the end of November. And it records that I was falling in love – that, too, in a thoroughly boyish manner and frame of mind.

The first entry is dated 10 October 1940.

10.10.40 A busy day, spent training the new and completely inexperienced Sergeant Pilots. 616 Squadron has deteriorated into a Flying Training School, or at best an OTU.

The day really started after release at 7 o'clock. JCD turned up; he is on sick leave after being mildly peppered by a Me 110. As usual when he visits the Squadron, the evening developed into a party. Much sherry, claret and kummel were consumed.

11.10.40 Woken up at 6.15, and after 'Begin the Beguine', 'J'Attendrai' and a stiff Enos managed to come to Readiness. At 8 o'clock Mac, John and I took off and had a tremendous triangular 'dog fight'. It started at 10,000 feet and ended up between the hangars; the troops were impressed, but the CO wasn't. It was colossal fun.

After breakfast John and I took a Maggie (Miles Magister two-seater trainer, used as a squadron 'run-about' aircraft) to Church Fenton, on a short visit to his American friends in the Eagle squadron. This is the 'Esquadrille Lafayette' of this war, but at the moment they have got nothing to fly except a Magister and a Link trainer. The original plan was to equip them with Brewster Buffaloes, but it looks as though these machines haven't really been turning up in any number, as Church Fenton hasn't had a smell of them. The Yanks were very pleased at the idea of getting Spitfires instead. (Shortly after this the Eagle Squadron joined us at Kirton.)

We arrived back at Kirton at about 12 o'clock and set off for home in the Lagonda. It was a race against time, as we had been told that there was grouse for luncheon. But it was a precarious race, as the Lagonda was suffering from odd noises in the transmission, which eventually came to a head on our way through Barnsley. The universal joint dropped out with a loud explosion. Mummy picked us up in the Rover, and we eventually reached and devoured the grouse.

After a game of squash at Wentworth, which John won 3–2, we went to Wortley to change and pick up Diana and Barbara. Dinner and dancing at the Grand. The evening ended with the customary roughhouse at Wortley.

12.10.40 Back to work again, via Doncaster and one of John Glover's aeroplanes. They have posted five more pilots away, which leaves us with four operational pilots in B Flight and only five in A. I don't suppose that they are deliberately trying to break the squadron's morale, but they are setting about it the best way.

16.10.40 Morale has ebbed pretty low today: I have more or less decided to give the squadron another month and then to apply for posting if things don't improve. I might just as well be an instructor at an OTU and I'm damned if I see why I should be anything of the kind.

26.10.40 Four day's leave should have begun today, but Flight Sergeant Burnard has been posted, so it was reduced to 48 hours. I am really sorry to see Burnard go, after working with him all through the summer and getting to know him and like him.

At Readiness all the morning and as luck would have it we were scrambled just as Sir Archibald Sinclair (Secretary of State for Air) and Leigh-Mallory were inspecting our Dispersal. Mac and I were off the ground in two minutes and I understand that they were duly impressed.

After lunch I steered a zig-zag course to Middle Wallop in the Maggie. John and 609 in general entertained me very well. It is as nice a squadron as I ever met and I really wish that my application to join it had not been so firmly quashed by the AOC. We had an amusing evening, including Rosalind Russell in 'That Girl Friday'.

29.10.40 A party in aid of ourselves, which ended up as my 'farewell party'. After much grousing on my part at our unfortunate and unpleasant fate as flying instructors, Jerry and the CO staged a phone call from Group posting one pilot to Biggin Hill. Burton asked me if I wanted to go and I had to say yes after my somewhat alcoholic invective against 616 and 'C' squadrons. The CO said 'Well if you want to go you bloody well can go and good riddance' which rather hurt my feelings and made me feel a bit sheepish. I went to bed still feeling sheepish and woke up feeling a damned fool as well. It wasn't until I had bathed and dressed that I learnt that I had been fooled. I sent a batman to the CO with a message that I had packed and was going early by car; then it was my turn to laugh. He came down the passage like a bullet to stop me!

1.11.40 On a battle climb this morning I could see the east and west coasts, the Hambledon hills, all Lincolnshire and the Wash, with the West Riding and Pennines below like a detailed plaster model.

Had an ugly dusk patrol. Through nearly 12,000 feet of storm cloud in the gathering gloom, came out at 500 feet and fortunately almost dead over Hornsea Mere. It was much too bad to try to get home in the dark, so felt for Catfoss, luckily saw the runway, and put down safe, though surprised!

My luck was in all round, as Catfoss mess happened to be throwing a cocktail party and I found myself in the midst of pretty gay and hospitable festivities.

4.11.40 At about four o'clock I set off for home and after a peculiarly cold and unpleasant drive, got there in time for a late tea. John, 'Aga' (Noel Agasarian, a friend of John in 609 Squadron) and Margaret Rawlings (distinguished actress and close friend of John) were there. Margaret just as one would expect from seeing her act – all the same deep inflexions and enthusiasms in her voice, slightly geared down.

We went to Wortley for Diana and Barbara and had an excellent evening at the Grand in Sheffield.

5.11.40 We had another dusk patrol in pretty unpleasant conditions. Gerry Jones was patrolling on the coast and intercepted a Heinkel 111. He shot it down, but received three bullets in the right arm from the Bosche rear gunner. In spite of his wound and the fact that his radio was shot away he got back in very bad weather conditions and carried out a perfect landing.

John and Aga turned up in a Maggie from Doncaster, just before dark, and another evening of devastation ensued. We went to bed in good order at 2.00 a.m!

11.11.40 John wrote to me saying that a 1,000 lb bomb had landed within 50 yards of his bedroom the other night; it may go off any time within the next 10 days – 'uneasy lies the head. . . .'

Odd though it seems, I think a lot about days when 1,000 lb time bombs, and incendiary bullets, and cannon shells, and all the rest of this nasty mess-up, will be things of the past. The dangers of forgetting the possibility of reverting to a normal life are enormous; it seems so incredibly remote. Two years ago, on this day and at this moment, I was driving with Alice between Cawthorne and Hillam; it was the night of the Badsworth Hunt Ball and we were en route for dinner with the Lyons. What, I wonder, will I be doing on Armistice Day 1942? Will that be as different from today as this is from two years ago? Will there be as much difference between 20 and 22 as there has been between 18, full of romanticism and illusions, and 20, as disillusioned as I suppose one can be at that age?

1.12.40. It's Sunday. On Friday I took 24 hours off to go to Wortley for a night. 'Buck' Casson lent me his car, as mine is still unserviceable, and I set off at about 2.30 p.m. for Cawthorne. When I arrived at tea time I had the biggest shock of the war – of my life, I suppose. Daddy had just received a telegram that John was missing after operations on the 28th. I went to Wortley in a daze. The truth hadn't really seeped through.

So it has happened at last. I suppose that it had to happen. I suppose that we were inordinately lucky to have survived intact as long as we did. Of course there is still hope. I tried all yesterday to phone Middle Wallop and, having at last got through, found that 609 was at Warmwell; by the time I had got through there everybody was in bed and asleep.

Poor Mummy! If John is killed – and I did my best to persuade her otherwise – I believe that even her brave heart will be broken. Now I realise what these last months have meant to her and that realisation is almost more bitter than the thought that John may be dead.

4.12.40 Thursday. Since my last entry I have been down to Warmwell. John's C.O., Michael Robinson, and all his squadron, were very kind and their sympathy was very obvious and very sincere; but they held out little hope.

On Tuesday I collected the Lagonda from Wallop, packed all John's personal belongings into the back and drove up as far as Stowe. The next day I came on up to Cawthorne.

Mummy and Daddy have both stood up to the shock remarkably well, though Mummy looks anything but well. Her bravery and balance are beyond all praise or expression.

Last night I dined at Wortley, and today I went to Wentworth for a game of squash with David Bethell.

My diary makes no reference to the circumstances in which John was shot down. I think it was quite a long time before all the facts were released by the authorities, though the general nature of the action was, of course, described to me by Michael Robinson, who was then commanding 609.

The squadron had been scrambled in the early afternoon of 28 November and ordered to climb to twenty-five thousand feet, south of the Isle of Wight. Robinson was leading the squadron. John's flight made up the second six of the twelve aircraft. A formation of enemy planes was reported to be approaching from the south and 609 was directed towards them by the controller. His directions were effective. A gaggle of Me 109s was spotted by the squadron, at the same height as themselves. But before they could be engaged they turned back and contact was lost. The controller ordered 609 to orbit and as they did so they gained some altitude. A few minutes later the 109s reappeared. They, too, had gained altitude and this time they had the height advantage. Some of them curved down to attack and Robinson ordered the squadron to break round towards them. Briefly, fleetingly, they clashed in the pale, cold, thin winter air.

John's voice called out, 'Whoopee! I've got a 109!'

It was the last ever heard of him.

The 109s had dived away, heading back across the Channel, and Robinson told his pilots to re-form. When they had done so John's plane was missing and he did not answer Robinson's repeated calls. After the squadron had landed one of the pilots reported having seen a torn parachute descending south of the Needles.

That was the account of the action given to me during my visit to Middle Wallop. What none of us knew until much later was that, shortly after the action, a signal was received by the RAF from the Luftwaffe stating that Oberst Helmuth Wieck, Germany's top-scoring fighter leader, had been shot down that afternoon off the Isle of Wight and asking whether there was any news of him. It had been the only contact between the two air forces that day. Wieck was the only German casualty, John and his number two the only British.

A few days later the body of a pilot, a torn parachute still attached to it, was washed ashore. It was the body of Helmuth Wieck. And so, in due course, we learned that John, in the last moments of a glorious life, had accounted for Wieck, whose tally of fifty-six aircraft destroyed in air combat had its origins in the Spanish Civil War. A few seconds later one of Wieck's companions had accounted for John.

It was the last encounter of the year, at squadron strength, between Spitfires and Messerschmitts. It certainly contributed nothing, one way or the other, to the course or outcome of the war. But it affected my life deeply. I think that hardly a day has gone by since then when I have not thought of John.

My diary records my feeling in the weeks which followed his loss.

23.12.40 After seventeen days I am back again at Cawthorne for Christmas. I don't remember much about those seventeen days, except that they have been a queer blur, a mixture of sharp unhappiness and restlessness. Unhappiness about John is gradually growing on me. For the first time I have felt a sickening desire to be away from aeroplanes and pilots: I suppose it will wear off. It's got to, if I'm going to be any good again.

25.12.40 Christmas Day. Last night I went to Wortley for dinner. It was the same as it always has been. I got there five minutes late, but five minutes before anybody came downstairs. Moss produces the sherry; it is always the same formula 'Good evening, Master Hughie; I think they're all upstairs changing yet, sir. I'll get the sherry, sir. Have you got everything you want, Master Hughie?' And then Archie stumps down, and Elfin, and Barbara and Diana, always late. And Diana looks even more beautiful than usual; and one half of me says you're in love, you fool, tell her, while the other half says you've no right to be in love with anybody. And tonight I almost did tell her and then stopped and felt a fool. But of course she knows and then I think she could be persuaded to love me. But then I am back full circle at the unaswerable argument that in my present occupation I can't make girls like Diana love me.

We went to church at 8 o'clock. No church bells, no row of candles on the High Altar; just an improvised table with two shaded candles, and a hidden congregation in the dark church: Holy Communion on Christ's birthday hiding secretly from German bombs. It is one thousand nine hundred years exactly since the first Supper Feast, and we have to celebrate like this, like Roman Christians in the Catacombs.

26.12.40 Boxing Day I went over to Wortley soon after breakfast. Barbara, Diana, Diana Booth, a KOSB officer and I set off for Wetherby races in Diana's Wolsey. It was a heavenly day; blue sky, pale and milky, south wind, sun low, watery, but with an incipient vestige of warmth. I was out of uniform, feeling like a conscientious objector in a sea of khaki, and thoroughly enjoying it. The last race was at 3 o'clock, and we were back at Wortley for a late tea. Champagne and oysters for dinner and the annual dance at Wentworth village, which must always be attended. Six pennyworth of slap and tickle but great fun. Cousin Maud was there giving the prizes, and Pickles and Brian, and the Landons, and, of course, busy little Hilda Talbot organising anything she could find to organise.

27.12.40 Back to Kirton in time for lunch. It looks as though we'll be moving back into the line fairly soon, which will be a good thing. The Harbour Master (slang for Commanding Officer) seems to think that we will go before the end of January.

1.1.41 Stap me, what a hangover! I fairly floated out of bed at lunch time. Whatever the new year brings, there was no lack of conviviality or good wishes at its inauguration.

14.2.41 After a long gap – brought about more by laziness than anything else – I have decided to start again. It has been an uneventful month, and on the whole a bloody one. We have had the annual messy and uncomfortable snow falls. There has been very little flying and most of the work has been dull routine. But I have had seven days of leave, from 3.2.41 to 10.2.41 and have been more and more weak-minded about affairs of the heart. It is becoming increasingly difficult to stop the rolling stone and I have more or less abandoned the effort. I adore Diana and that's the whole story. If things develop, and I get shot down, it will be just too bad; but there it is.

16.2.41 Yesterday was a fairly busy day in the air. The Squadron did a convoy patrol in the morning. A Dornier 215 flew straight through us and was not intercepted, which must have looked pretty black to the Navy if they saw it. We came back through very bad weather, having to split into sections. 71 (Eagle) Squadron had a section going out to relieve us, led by 'Shortie', who got mixed up in a cloud and flew into the sea – a very serious loss for them. ('Shortie' Keough was one of the gallant American pilots who had found ways of getting themselves into the RAF and into fighter squadrons even before the Eagle Squadron was formed. He had been with John in 609 Squadron during the Battle of Britain.)

Later I did another convoy patrol with 'A' section and at 7 o'clock an hour's night flying in no-moon conditions. Today has been very ordinary. We did a flight climb and exercise first thing this morning, which proved fairly hectic. A long sleepy afternoon produced no scrambles and the day finished with a glass of beer at the Red Lion with Peter Clapham, who goes to Church Fenton tomorrow.

We are by the way of taking off for Wittering at 8 o'clock tomorrow for duty wing, but it doesn't look as though the weather will be fit. If we go, I am taking 'A' Flight. It will be odd to be on the real thing again after the winter months of inactivity. I only hope I don't make a clot of myself and get shot up on our first patrol over the other side. With my usual procrastination I have made no plans or preparations for being a prisoner of war.

P.S. Mac and Sergeant Arnold intercepted a Ju 88 out at sea this morning. They both squirted, but it escaped in cloud.

That was the last diary entry written at Kirton and it reflects the fact that our days as a 'C' Squadron were over. We were settling down as a team and preparing again for action. Of the old brigade there were, apart from myself, Ken Holden, 'Buck' Casson, John Brewster, all survivors from the pre-war days at Doncaster, and Roy Marples who had been with us since the beginning of 1940. Ken had taken over as commander of 'A' Flight after Jerry Jones's night-time encounter with the rear-gunner of a Heinkel bomber. Colin Macfie had become a well-liked and respected commander of 'B' Flight.

It is strange that the diary does not mention two pilot officers who had joined soon after we got to Kirton. They were James (Johnnie) Johnson and Philip Heppel. Both were to make their marks strongly in the months to come and Johnson in particular was to emerge as the outstanding fighter pilot on the allied side in the war. They were very different types from each other on the surface, and different from me also. Yet we became close friends.

Johnnie, the son of a Leicestershire policeman, was as tough and rugged as they come, he had the strongest constitution of any man I ever knew. His language was more expressive than correct. He was an out-and-out extrovert, ready for anything at any time of the day or night. Heppell – commonly known to one and all as 'Nipple' or 'Nip' – was very young, slightly built, though tall and – misleadingly – rather soft looking. He was a member of a long-established Northumberland business family and as such was to have a distinguished commercial career after the war. You would not have thought, when he joined us, that he could say boo to a goose. But he was an above-average pilot, an intelligent man and an exceptionally brave one.

In the last week of February rumours of our impending move were confirmed. We were ordered to Tangmere, to relieve 66 Squadron. Tangmere! It was the best we could have hoped for, a station with a splendid tradition. It was not only a station to be proud of, but one to enjoy from the point of view of its position, lying at the foot of the South Downs three miles east of Chichester and just south of Goodwood. We flew there on 26 February. From the account which I wrote later that year of our last night at Kirton, I imagine that some of us may have felt a little queasy on the way:

> It was a good party, as parties go.
> The Station Commander, Stephen Hardy, was also about to leave and a farewell dinner both to him and to us was arranged. We had wintered at Kirton quietly and happily, licking our old wounds and building a new squadron on the honourable ruins of the old. And so, though I think everyone was pleased to be moving south again, there was a definite twinge of sorrow at leaving behind so many good friends.
> However, nobody entering the mess that night at eleven o'clock would have discerned any trace of mourning. Where all the champagne came from and who bought it I never found out. Arrive it did, in a never-ending stream. I have to admit that for me the evening is more an impression than a memory. I seem to see Lord Tichfield (Honourary Air Commodore of 616 Squadron, later Duke of Portland) sitting happily in a corner, glass slightly awry, blinking and beaming benevolently at the riotous company; Johnnie, half-lying in a chair, solemnly pouring a bottle of champagne down his collar, having presumably reached capacity internally; Nip, a bottle tucked under each arm, wandering around the

room with a more than usually dazed expression on his face; Ken, his great round face glowing like an evening sun, with a perpetually replenished pint pot of black velvet, swaying slightly in the middle of the room and dispensing fatherly advice to anyone who cared to listen.

Yes, Kirton gave us the right kind of send-off. In retrospect it seems altogether fitting that we should have made merry on the eve of our departure. Spring was in the air; and by the time the leaves fell again there would be few enough of that company left to enjoy champagne.

CHAPTER EIGHT

Conflict over France

Tangmere had been badly bombed during 1940. It was bombed again on several occasions in the weeks following our arrival. No doubt it was a nice easy target for the German bombers to find – close to the sea and surrounded by several good identification points. Also there was a night fighter squadron on the station, Number 219, which at that time was operating Blenheims but was soon to re-equip with Beaufighters carrying the new airborne radar. German intruders would hang around until one of the Blenheims came in to land and some lights were turned on. Then things used to go bang in the night.

Extraordinarily little damage was done. The raids had more nuisance value than anything else. The nearest thing to utter disaster came one evening when a bomb landed hard by the officers' mess during the pre-dinner drinking hour when the place was full of people. A few yards one way or another and that bomb could have disposed of a fair proportion of Tangmere's pilot strength. As it was, a large quantity of good liquor went west, for the bomb exploded alongside the barman's store-room, part of the back quarters of the mess collapsed in a pile of rubble and many window panes were shattered. In the crowded ante-room we all nearly died of fright. But casualties were mercifully light.

As a result of all this, dispersal was the order of the day. The station officers' mess was moved out to Shopwycke, a large country house between Tangmere and Chichester, which became the home of a well-known preparatory school for boys after the war. We moved our planes to Westhampnett, a satellite field at the foot of the Downs below Goodwood racecourse; and we went to sleep in a charming little house called 'Rushman's,' alongside Oving Church. One of the other Spitfire squadrons, Number 610, was already operating from Westhampnett and its pilots lived in a farm house on the north-east edge of the field.

The third Spitfire squadron – Number 145 – operated from another satellite field, about five miles to the south. The Sector Operations room had already been re-established in a requisitioned college building on the outskirts of Chichester.

No one could say that dispersal in the Tangmere sector had not been carried out to the ultimate possible extent. Perhaps the Germans got to hear about it, for quite soon after all these moves had been made, the regular night raids died out.

We were happy from the start at Tangmere, and things moved quietly enough to begin with. We took our turn at readiness and carried out convoy patrols. When the moon was high we were sent up on night-flying patrols. The planners at Fighter Command dreamed up a new way of using day fighters at night. It was called 'Layer Patrol', because it consisted of a number of fighter aircraft flying up and down the same patrol line, each at a different, pre-set altitude. The idea was that, on a bright, moonlight night, there would be a fair chance of spotting at least some of the considerable force of enemy bombers flying through the patrol line.

My diary, written at the time, records one such operation and the impact it had on me.

On the 10th March we were called in to reinforce the night fighters during a heavy attack on Portsmouth. We took off soon after 10 o'clock to patrol in layers, aircraft at 1,000 feet intervals, from 13,000 to 23,000 feet. It was my first experience of watching a night blitz from the air and I am glad I did not miss it. I suppose that the night-fighter and bomber boys get used to it, but for me it was one of the most unreal yet ghastly spectacles I have ever seen.

Looking down from 21,000 feet on that brilliant moonlit night, Portsmouth seemed to be a great lake of running liquid fire – constant explosions followed by huge dull-red glows; constant eerie snakes of tracer uncoiling up to meet me; flares hanging apparently motionless in the sky, flares in their twenties and thirties, lighting up great areas of dockland or town when they landed; huge squares of incredible brilliance as baskets of incendiaries found the mark, burning fiercely bright, like a million pinpricks of hot metal. It seemed to me impossible that anything should live or stand in that fantastic and horrible mess and I remember thanking God that I was not a bomber boy. For the first time I began to feel really angry. I would have sold my soul to find a Hun up there, to stalk behind and underneath, to give him the works up his dirty belly to watch him fall, another ball of fire.

It seemed incredible that, with all that fire beneath and the brilliant moon above, I should see nothing. The air must have been stiff with aeroplanes. Up and down from Selsey to the Needles, hood open, cold as hell, searching the darkness, probing the night for the bombers which were certainly there but apparently invisible. God, what a shambles! God, what a hell! God, what a bloody war! God damn and blast the bloody Germans!

'Hello, Rusty 17, Beetle calling. Return to base. Is this understood? Over.'

'O.K. Beetle, O.K., O.K. Rusty 17 returning to base. Listening out.'

I came down in a wide sweeping dive around the Isle of Wight, still searching, still bloody angry, but not displeased with the idea of landing and getting some liquid inside me. No trouble about finding the aerodrome with that moon – the Chichester by-pass stood out like a curved white ribbon. Speed down; wheels; fine pitch; a bit high – throttle right back and flaps; blast these goddam exhaust flames; steady turn onto the concealed glim lamps; 100 m.p.h. and a bit of engine; watch that sector light – no flood; it's coming up to meet us pretty fast now; throttle right back – there goes the hedge – feel for it – Christ, it seems black enough now and going like a cat in hell; bump she goes – keep straight, you bitch! Rudder, rudder, brake! Not bad, but not good. It feels colder still down here. Hope they've got the bar open.

'See anything, Sir?'

'Yes, a bloody sight too much.'

I did not re-read those words until some forty-six years after they were written. For what it is worth they are the authentic record of how one 20-year-old fighter pilot saw and remembered a blitz on Portsmouth all those years ago.

The new C-in-C of Fighter Command, Air Marshal Sir William Sholto Douglas, had plans for the day fighter squadrons in the south of England that summer in which night operations counted for little. The massively-built air marshal, who had been a fine fighter pilot himself in the First World War, was thinking offensively. Across the Channel the airfields were still thick with German fighters and it was Sholto Douglas's intention to bring those planes to battle. But not over Britain. Carry the battle to the enemy. That was his plan. It was a project which commended itself strongly to Leigh-Mallory, now commanding 11 Group. He set about reorganizing his command in preparation for an early opening of the offensive.

His first step was to establish a new post at each sector station – the post of 'Wing Commander Flying'. Previously each individual squadron commander had been responsible direct to the station commander, who in turn was responsible to the sector commander. (Sometimes, indeed quite often, the station commander was also sector commander.) If more than one squadron flew together as a formation, the senior squadron commander would lead. Now the new wing commanders flying – or wing leaders, as they came to be called – assumed responsibility for co-ordinating all flying activities and for leading the Wing formations.

To Tangmere, for these duties, came Douglas Bader. And at about the same time Group Captain 'Woodie' Woodhall was appointed sector commander. The old Duxford firm was back in business.

I had been transferred to 'A' Flight as second in command of the flight, under Ken Holden. By an extraordinary stroke of luck Bader elected not only to fly with our squadron but to put his Spitfire in the charge of the 'A' Flight. So there began a close personal association between us which lasted throughout the summer until that dreadful day in August when Bader did not return from a fight. Whenever he flew, I flew. We did some sixty offensive sweeps over France together, most of them compressed into a few hectic weeks of high summer.

Our partnership nearly ended on the very day it began. We sat up late in the mess one evening, several of us together, discussing battle formations. We were expressing dissatisfaction with the formations adopted in the past. In 1940 the squadrons of Fighter Command had all flown the same basic formation. The twelve aircraft of a squadron were divided into four sections of three aircraft each. Normally these sections flew in 'Vic' formation – that is to say one plane on each side of the leader and a little behind him. When going into action this formation would be changed to line astern. The last man in the last section was responsible for 'weaving', swerving from side to side in order to keep a good look-out to the rear.

Having frequently flown in that tail-end position, I knew well the difficulties and hazards involved. If you weaved too much, you got left behind. If you did not weave enough, you got picked off. I said that I thought it was a lousy formation and that no variation of it would be any good. For instance, we had been experimenting with the idea of flying in three sections of four, aircraft line astern. The last man in each section was still excessively vulnerable. We needed to find something quite different, some way of flying which would cut out all that weaving around and enable everyone to cover everyone else's tail.

The half-pints went down again and again while we argued the toss. I was in favour of trying line-abreast formation, already extensively used by the Germans. I argued that four aircraft flying side by side, each one about fifty yards from his neighbour, could never be surprised from behind. The two on the left would cover the tails of the two on the right, and vice versa. No enemy plane could get within shooting distance without being seen. If attacked, you would break outwards, one pair to port, the other pair to starboard.

Next morning I was sitting at breakfast feeling just a little queasy and concentrating on black coffee when Bader came and sat at the same table. He never drank beer, or anything else containing alcohol, and he appeared quite the reverse of queasy. Bright, breezy and aggressive are the words which came to mind in describing his demeanour. I did not pay too much attention when he referred to our discussion of the previous evening. I just went on concentrating on the coffee. When he said that he had been thinking about the idea of line abreast formation and had decided to try it, I let the matter pass. His next remark however, riveted my attention.

'We'll give it a try this morning – make a pass down the Pas de Calais. Probably find something there. You'd better come with me, Cocky. I'll fix it with Woodie (the Station Commander) and let you know what time.'

He told Paddy Woodhouse, the CO of 610 Squadron, who was sitting with us, to get himself a number two and to stand by to come along too. Then he stumped off to make the arrangements.

'Oh God,' I thought, 'you bloody fool, Dundas. That will teach you to keep your big mouth shut.' I pushed my coffee cup away and left the room feeling queasier than ever.

We flew east along the coast, climbing steadily. I was on Bader's left, Woodhouse on his right. To the right again was Woodhouse's number two, a sergeant pilot of 610 Squadron. The weather was fine, the sky was clear. We maintained R/T silence, as arranged. We crossed the North Foreland at about twenty-five thousand feet and headed on towards Dunkirk. Wisps of white condensation began to stream out behind us. Bader lost a little height to drop beneath the revealing vapour-trail layer. Approaching the French coast we turned through 180 degrees to starboard and straightened out on a westerly course which would take us just clear of Calais. The R/T crackled. Woodhall's voice came through, deep and resonant.

'Hello Dogsbody, there are some customers in your area. Quite close to you, about the same height.'

'O.K. Woodie. Watch it boys.'

Bader flew on, straight and level. I started twitching. Still, the formation seemed foolproof. Certainly nothing could get close to Woodhouse or his number two without my seeing it. I hoped they were scanning the sky behind me with equal intensity. I saw a wisp of white vapour form for a second, three or four miles behind and a little above. Something had flown momentarily into the condensation layer. Then my eyes focused on the right place and distance. There they were – half-a-dozen 109s curving round and down to fall in behind us, like hounds on the trail.

'Bandits behind us Dogsbody. Five o'clock. A little above. About three miles.'

A short pause.

'All right, Cocky, I see them. Now wait for it, everybody, wait for it!'

We flew on as though nothing was happening. My stomach turned somersaults and the sweat broke out in pin-pricks all over. Quick check to make sure that reflector sight was on and gun-button turned to 'fire'.

'O.K. boys, get ready for it . . . now BREAK.'

We whirled round in a turn so sharp that the haze of incipient black-out almost shut off the sight of Bader's tail wheel as I followed him round. Straightening out on our original course I peered around for the 109s. According to my masterly exposition of the previous night they should now be in front of us, waiting to be shot down. Where were they? Where the hell were they?

Again the explosions were utterly shattering. Again the cockpit filled with the acrid white smoke of hot glycol fumes escaping from a punctured coolant tank, so I could see nothing at all as I hauled the plane round in violent evasion. But this time I had a jettisoning system for my hood. I pulled the little rubber ball and banged the perspex upwards and outwards with my forearm. The hood flew off and the smoke streamed out of the cockpit and I could see again. I was in a steep diving turn. Not another plane to be seen. I turned north and eased the dive, throttling right back. Already the temperature was going off the clock. But I still had over ten thousand feet on my altimeter and Hawkinge, the little airfield behind Dover, was within easy gliding distance. Try the radio . . . nothing doing. A Spitfire dived past then swooped up close to me. It was Paddy Woodhouse. He tapped his R/T mask. I shook my head, to show him that I could not receive.

Now to get down intact. I was over the coast, committed to a landing.

Hawkinge was a small airfield, so there was still plenty of opportunity for breaking my neck, if I misjudged the approach. I pushed the throttle forward to see whether I had any engine left at all. The noise and vibration were so horrific that I quickly pulled it back, cut the magneto switches and turned off the petrol. Better to overshoot than undershoot. An undershoot would mean curtains for sure, so I came in deliberately high, from the west, and when I selected 'flaps down', one flap came down and the other stayed up, so I quickly selected 'flaps up' again. God Almighty! Without flaps I was not just too high, I was a bloody sight too high. Still, I could not go round again. I had to get onto the ground before reaching the other side of the airfield, or I was for it.

I side-slipped violently and let the nose drop. I finally got down to ground level by the time I was about two-thirds of the way across the field. But I had much too much speed on. It must look more like a beat-up than a landing. Nothing for it now but to stuff her into the ground. I did just that. There was a horribly expensive noise of tearing metal as we churned along in a cloud of dirt and flying stones and charged through a lot of Spitfires, parked out at their dispersal point, eventually coming to rest a few yard from a pilot's hut.

Paddy Green, an Auxiliary Air Force officer who had been in 601 Squadron before the war and who now commanded the Spitfires at Hawkinge, came out to see what was going on. His planes were all brand new, specially hotted-up for high-speed, low-level reconnaissance work. He looked anxiously round to see how many had suffered as a result of my precipitous arrival. Astonishingly, they were all intact. Paddy walked over to me. I was still sitting in my cockpit, dazed and a bit surprised to find myself still in one piece. He stepped onto the wing and greeted me with the cheerful observation that with a little more effort I could have written off his whole bloody squadron. I said to hell with his horrible aeroplanes and how quickly could he get me close to a large, strong drink?

About two hours and half-a-bottle later Ken Holden turned up with the Magister to fly me back to Tangmere. He told me that Bader had definitely shot down one Messerschmitt and claimed another probably destroyed. Woodhouse also claimed a probable. That afternoon we had a post-mortem. We all agreed that the main advantage of the new formation had been proved. It was practically impossible to be taken by surprise from behind. Our mistake seemed to have been we had mistimed our break, so that one or more of the enemy planes was still behind us when we straightened out.

And so, in spite of my disagreeable experience, we decided to press on with the new formation. We called it 'finger fours' because the planes of each section were positioned in the same relative way as the tips of the outstretched fingers of one hand. Bader immediately started practising with the whole Wing of three squadrons flying together in this way and it became popular with all pilots. A number of other Wings adopted the same idea, though there were exceptions. The Biggin Hill Wing, for instance, led by 'Sailor' Malan, continued to fly with their aircraft in line astern. Malan was never converted and there were many arguments, over the years between those who supported his tactics and those who adopted fingers four. For myself, I was a militant champion of the new method and stuck to it unwaveringly until the end of the war.

Nearly forty years later I received a letter which provides a curious postscript to my account of that adventure. It came to me from Mr Denis Knight, who had written a book called *Harvest of Messerschmitts* – a record, culled from the 1940 pocket diary of the daughter of a Kent village postmaster, of all the planes which crashed within the boundaries of the village of Elham during the summer and autumn of that year. My own inglorious descent on 22 August had contributed to the score. Anyway, Mr Knight had written to ask me for some information about Hawkinge and in my reply I had told him of my crashlanding there. On 15 May 1980 he wrote to me and told me something which made me realize just how lucky I had been not to have died that day. He said:

'Following your letter telling me about your mishap at Hawkinge, I have discovered something quite extraordinary and rather funny! On the day that Douglas Bader and yourself stooged about off Dover to test out your theory on operating two pairs of Spitfires in line abreast an exalted exponent of aerial combat who'd thought of the same idea two years earlier was also off Dover. As you know the finger-four formation was devised by Werner Moelders and first used by JG51, before it was adopted by all German fighter units. Oberst Werner Moelders, then the highest scoring Luftwaffe pilot, led his Staff Flight over the Channel on 8 May 1941. He filed combat mission report No.292 – shooting down his 68th victory, a Spitfire at Dover!'

'Rather funny?' I certainly would not have thought so, at the time.

The offensive operations planned by Sholto Douglas and Leigh-Mallory began in earnest in mid-June. From then until the end of the summer we flew over France almost every day, except when the weather was bad. Often there were two sweeps in one day, occasionally three.

The Luftwaffe reacted as our Commander-in-Chief had hoped. Day after day the Messerschmitts swarmed up to give battle. Night after night, in the bar at Shopwycke House, there was a wake for missing friends.

There was a close bond between the three Spitfire squadrons at Tangmere that summer. Bader welded the Wing into a single unit and we all knew each other well, so that the losses sustained by the other squadrons were almost as painful as our own. Ken Holden was posted in May to command 610 Squadron on the other side of our field at Westhampnett, and Bader had brought in Stan Turner, a Canadian who had flown with him during the battle in 1940, to command 145. It would be difficult to imagine four men more outwardly dissimilar than Bader and his three squadron commanders, Burton, Holden and Turner. You would never have thought that the punctilious Burton, indelibly stamped with the Cranwell hallmark, could have welded easily or happily with the wild-eyed, hard-drinking Stan Turner, who affected to care nothing for discipline and was trenchantly irreverent about all who were set up in authority over him. You might have thought that Holden, the stolid, steady Yorkshireman who liked all things to be well-ordered, would have been incompatible with the mercurial, impatient, unconventional Bader. Yet these four men in fact formed an extraordinary harmonious and effective leadership team. They admired and understood each other. They were as one on the ground and in the air.

My memory of that summer is one of sharp contrasts; of the pleasure of being alive and with friends in the gentle Sussex summer evenings; of visits from Diana, when we would dine and dance in Brighton, or sit long on the balcony outside the Old Ship Club at Bosham watching the moon on the water and listening to the tide lapping against the wall beneath us and memories of tearing terror when, at the end of a dogfight, I found myself alone with fifty miles of hostile sky between me and the Channel coast and the hungry 109s curving in to pick off the straggler.

On some faded sheets of paper there has survived a description which I wrote, at that time, of an afternoon and evening which could have been one of many. It is incomplete, a fragment, not part of a diary or larger chronicle. I cannot remember writing it; I do not know exactly when I wrote it, for it is undated. Nor do I know whether it ever had an ending or whether perhaps I just got tired of writing and went to bed. But such as it is, it brings back sharply the feel and taste of those far-off days when I was very young and just discovering life and death stretched out its hand to touch me every day. I quote it, just as it was written then:

It was hot in the garden, lying face down on the lawn, a pot of iced shandy by my hand, Robin (my golden retriever) huffing and puffing and panting at the ants. Odd to be lying there peacefully, listening to the click of croquet balls, the blur of voices, the gramophone. The shandy sharp, cold, stimulating.

'Hullo, Cocky.'

'Hullo, Johnnie.'

'Get a squirt this morning, Cocky?'

'Yes, Johnnie, I got a squirt. Missed the bastard as usual, though.'

'Another show this afternoon, Cocky. Take off 15.30.'

'Yes, I know; take off 15.30.' Three hours ago, over Lille. It happened yesterday, and last week, and last month. It will happen again in exactly two and a half hours, and tomorrow, and next month.

The grass smelt sweet in the garden, and the shandy was good, and Robin's panting, and the gramophone playing 'Momma may I go out dancing – yes, my darling daughter.'

It was hot at dispersal and the grass, what was left of it, brown and oil-stained. The Spitfires creaked and twanged in the sun.

'Everything under control, Hally?' (Flying Officer Hall was the squadron engineer officer.)

'Yes, Cocky, everything under control. DB's not ready yet, but it will be.' (DB were the indentification letters of Bader's plane.)

'Well, for Christ's sake see that it is, or there'll be some laughing off to do.'

'It will be ready, Cocky.'

'O.K., Hally.'

Inside is as hot as outside. The pilots, dressed almost as they like, lie about sweating.

'Chalk please, Durham.'

They all watch as I chalk initials under the diagram of twelve aircraft in three sections of four. Nobody moves much until I have finished and written the time of take-off.

'Smith, you'll be with D.B. Nip, you and I on his right. Johnnie, you with the C.O. and two of 'B' Flight. O.K.?'

'O.K., Cocky.'

Here comes D.B.

'Why the bloody hell isn't my aircraft ready? Cocky, my bloody aeroplane's not ready. We take off in 20 minutes. Where's that prick Hally?'

'It's O.K., D.B., it'll be ready. I've seen Hally.'

'Well, look at the bloody thing. They haven't even got the cowlings on yet. Oi, Hally, come here!'

Christ, I wish we could get going.

'Chewing gum, Johnnie, please. Thanks pal.'

'O.K., D.B.?'

'Yes, Cocky, it's going to be O.K.'

We walk together again, as far as the road.

'Well, good luck Cocky. And watch my tail, you old bastard.'

'I'll do that D.B. Good luck.'

Just time for two or three more puffs before climbing into A for Apple.

'Everything O.K., Goodlad?' (the fitter who looked after my plane).

'O.K., sir.'

'Good show. Bloody hot.'

Climbing in, the hottest thing of all. The old girl shimmers like an oven, twangs and creaks.

'Good luck, sir.'

'Thanks.'

Up the line D.B.'s motor starts. 610 have formed up and are beginning to move off across the airfield as we taxi out – D.B., myself, Smithie, Nip, then two composite sections from both flights.

Straggle over the grandstand at Goodwood in a right-hand turn and set course east in a steady climb, Ken's twelve a little above and behind to the left, Stan's out to the right. Ten thousand feet over Shoreham. The old familiar, nostalgic taste in the mouth. Brighton – Maxim's last Saturday night; dancing with Diana in the Norfolk. Beachy, once a soft summer play-ground, now gaunt buttress sticking its chin bluntly out towards our enemies. Spread out now into wide semi-independent fours. Glint of perspex way out and above to the south shows Stan and his boys nicely placed between us and the sun. Dungeness slides slowly past to port and we still climb steadily, straight on, way out in front.

Twenty-five thousand.

'Levelling out.'

Puffs of black ten thousand feet below show where the bombers are crossing between Boulogne and le Touquet. Six big cigars with tiers of protective fighters milling above them.

'Hello, Douglas, Woody calling. There are fifty plus gaining height to the east.'

'O.K. Woody.'

'Put your corks in, boys.' Stan.

Over the coast at Hardelot we nose ahead without altering course.

'D.B., there's some stuff at three o'clock, climbing round to the south west.'

'O.K., I see it. Stan, you deal with them if necessary.'

'O.K., O.K. Don't get excited.'

Usual remarks. Usual shouts of warning. Usual bad language. Usual bloody Huns climbing round the usual bloody way.

St Omer on the left. We fly on, straight and steady in our fours, towards Lille. Stan's voice:

'They're behind us, Walker squadron. Stand by to break.'

Then: 'Look out, Walker. Breaking starboard.'

Looking over my shoulder to the right and above I see the specks and glints which are Stan's planes break up into the fight, a quick impression of machines diving, climbing, gyrating. Stan, Ian, Tony, Derek and the rest of them are fighting for their lives up there.

Close to the target area now. More black puffs below show where the bombers are running in through the flak.

'Billy here, D.B. There's a lot of stuff coming round at three o'clock, slightly above.'

Quick look to the right. Where the hell? Christ, yes! There they are, the sods. A typical long, fast, climbing straggle of 109s.

'More below, D.B., to port.'

'O.K., going down. Ken, watch those buggers behind.'

'O.K., D.B.'

'Come on, Cocky.'

Down after D.B. The Huns are climbing fast to the south. Have to get in quick before those sods up above get at us. Turn right, open up slightly. We are diving to two or three hundred feet below their level. D.B. goes for the one on the left. Nipple is on my right. Johnnie slides across beyond him. Getting in range now. Wait for it, wait for D.B. and open up all together. 250 yards . . . 200 . . . wish to Christ I felt safer behind . . . 150. D.B. opens up. I pull my nose up slightly to put the dot a little ahead of his orange spinner. Hold it and squeeze, cannon and machine guns together . . . correct slightly . . . you're hitting the bastard . . . wisp of smoke.

'BREAK, Rusty squadron, for Christ's sake BREAK!'

Stick hard over and back into tummy, peak revs and haul her round. Tracers curl past . . . orange nose impression not forty yards off . . . slacken turn for a second . . . hell of a mêlée . . . better keep turning, keep turning, keep turning.

There's a chance, now. Ease off, nose up, give her two length's lead and fire. Now break, don't hang around, break! Tracers again . . . a huge orange spinner and three little tongues of flame spitting at me for a second in a semi-head-on attack. Round, round, so that she judders and nearly spins. Then they're all gone, gone as usual as suddenly as they came.

'Cocky, where the hell are you? Are you with me, Cocky?'

There he is, I think. Lucky to find him after that shambles.

'O.K., D.B., coming up on your starboard now.'

'Right behind you, Cocky.' That's Johnnie calling.

'O.K. Johnnie, I see you.'

Good show; the old firm's still together.

It was cooler, on the lawn, and still. The shadows from the tall trees stretched out to the east. Robin lay beside me pressing his muzzle into the grass, huffing at insects. The pint pot of Pimms was cool in my hands and the ice clinked when I moved. The cucumber out of the drink was good and cold and sharp when I sucked it.

'Hullo Cocky.'

'What-ho Johnnie.'

'Tough about Derek.'

'Yes, Johnnie; and Mab.'

The croquet balls sounded loud to my ear, pressed in the grass. The distant gramophone started again on 'Momma, may I go out dancing'.

'Come on, you old bastard, let's drink up and get out of here.'

The tide washed up the creek to Bosham and splashed against the balcony of the Old Ship. We sat and sipped our good, warm, heartening brandy and watched the red sun dip through the western haze, watched the stars light one by one, watched the two swans gliding past like ghost ships.

'Cocky.'

'Yes, Johnnie.'

'Readiness at four a.m.'

'O.K. let's go.'

That was the way of it at Tangmere in high summer 1941. That, word for word, is how I wrote it down, in some moment of self-release on eleven sheets of pale blue writing paper which then lay unregarded among other old papers for twenty years.

The losses were grievous during those weeks of maximum effort. In the few weeks between 20 June and 10 August the squadron lost twelve pilots – more than half of its full establishment. But in the same period we claimed twenty-one enemy planes definitely destroyed, twelve probably destroyed and twenty-one damaged. One loss affected me particularly strongly – that of John Brewster, one of the last remaining auxiliary officers who had joined the squadron with me at Doncaster two years before. That left only Buck Casson and myself of the old originals, though of course Ken Holden was still there not far away, across the field with 610 Squadron. But morale was sky-high, despite the sadness of constant losses, and there was none of the shattering of the spirit which had accompanied the squadron's collapse the previous August.

There were many things to laugh at in the midst of the tiredness and the fear. I remember Ken Holden in the mess one evening after coming home from a sweep in which the strap of his goggles had been severed at the back of his head by a bullet. No one could have a closer shave than that and for once Ken's stolid and monumental calm was shattered. His incredulous description of the incident, constantly repeated, was an unintentional music hall turn and all poor Ken got in place of sympathy was an endless howl of laughter which became more and more hysterical as he went on and on about it.

Douglas Bader had been one of the most ribald members of the audience. But a day or two later Ken got his own back. Bader found himself alone with only one other Spitfire after a dogfight over France. There were many 109s about and to make matters worse Bader had lost a lot of altitude in the fight. He was in a dangerous situation. Bader called Ken on the radio and asked where he was. Ken replied that he was at twenty thousand feet and gave his position. It was approximately the same as Bader's, though he was some fifteen thousand feet lower.

'Can you come down and join me?' Bader called.

There was a moment of silence, broken by Ken's deliberate and unruffled Yorkshire accent.

'Nay come up and join us,' said Ken. And there the matter rested.

We lost both our flight commanders within a week, early in July. Colin Macfie was first to go, but fortunately we soon heard that he had been taken prisoner. The other flight commander at that time was a man called Gibbs, a regular oficer and considerably older than the rest of us. He had been posted in when Holden moved over to 610 squadron. 'Gibbo' was a pilot of exceptional skill, particularly at aerobatics. But he had no operational experience and, being set in his ways, it became evident to me that he could not last long. Sure enough, on 9 July he failed to return from a sweep. Subsequently we learned how he saved his life by having recourse to his extraordinary aerobatic skill.

After his Spitfire had been hit and damaged he had come down to a fairly low altitude when his engine failed completely. Gibbo evidently decided that a forced landing would give him a better chance of escaping capture than taking to his parachute. He selected a suitable field and headed down. At this point he found he had company in the form of 109s with unfriendly intentions. With no engine, and little altitude for manoeuvre, he was a sitting duck – or so it must have seemed to the Messerschmitt pilots as they curved in to administer the *coup de grâce*. However Gibbo had a trick left in his hand. Though only about two hundred feet above the ground, he inverted his Spitfire and continued his glide in the upside-down position, rightly guessing that the German pilots would assume that he was about to crash and would just sit back to watch. This is exactly what they did and I can imagine their surprise when, at the very last second, Gibbo rolled back the right way up plopped his Spitfire down on its belly and scrambled unhurt from the cockpit. He then took to his heels and never stopped running until he reached the Spanish border and safety.

It was Gibbo's disappearance from the scene which resulted in my elevation to flight commander. I was unashamedly proud of achieving this position a fortnight before my twenty-first birthday. And something still better was to follow. On 16 July I sat for my portrait by Captain Cuthbert

Orde, who was at that time going round the fighter squadrons doing portraits commissioned by the Air Ministry. I enjoyed my morning with Orde, one of the kindest and friendliest of men as well as a fine artist. But I was rather surprised and embarrassed when, at the end of it, he said:

'I've left room for the DFC. The people I draw always seem to get the DFC.'

Four days later his prophecy was fulfilled. I was able to telegraph my parents the glad news on the eve of my coming of age. Apart from Gillam, who had already left us when his DFC was gazetted, I was the first member of 616 Squadron to be honoured in this way. There were others who had done more to earn it. But that consideration did not detract from the riotous celebration which followed.

CHAPTER NINE

A Leader Falls

The eighth day of August dawned with broken cloud at medium height and a clear fine sky above. The Tangmere Wing was instructed to provide high cover for a mid-morning raid in the Bethune area.

Things went wrong from the start. As we climbed away down the coast, Ken Holden quickly clicked into position on our port side with 610 squadron. But where was 41 Squadron? This unit had recently replaced Stan Turner's 145 Squadron, which had been sent north for a rest. Now it was nowhere to be seen. It had somehow missed the rendezvous and the strict rule which ordained absolute radio silence in the preliminary, build-up stages of a sweep prevented Bader from calling to find out where it had got to.

As we began to gain height, Bader waggled his wings insistently – a signal indicating that for some reason he wanted me to get close beside him. I slid across and tucked my wing in two or three feet from his. From his gesticulation I understood that he wanted me to go ahead. Then he spoke two words: 'Airspeed indicator'. I realized that his own instrument was unserviceable and that he wanted me to take the lead in order to ensure that we climbed at the right speed to bring the Wing over the target area at the exact time planned. Both the effectiveness and the safety of each constituent part of the British force depended on the observance of accurate timing by all the others. Already the Tangmere Wing was below strength, owing to the failure of 41 Squadron to rendezvous. At all costs we must now ensure that we got to the right place at the right time.

As always I had a note written on the back of my left hand in indelible pencil to remind me of the time we were due over the French coast and the speed we must make on the way. I slid out and nosed ahead, then settled down to concentrate on the job. The exact height flown by the high-cover Wings was left to the discretion of leaders, depending on the conditions they found prevailing. The sun was bright and brilliant, unveiled by any layer of high haze or cirrus cloud. The puff-balls of white cumulus below provided a background which would show up the outline of any planes seen from above. I knew that Bader would wish to go in high on such a day and I adjusted my throttle setting and rate of climb accordingly.

We were flying at about twenty-eight thousand feet as we approached the French coast between Boulogne and le Touquet. Ken Holden had taken his squadron across to starboard and was about two thousand feet above, between us and the sun. We crossed on the dot of the appointed time. Bader levelled off and went ahead, calling on the radio-telephone to tell me he was resuming the lead.

Someone reported 109s to port and below, a little ahead. There they were, showing up clearly, a gaggle of several enemy planes, climbing in line abreast formation. I smelt a rat and I think Bader did, too, for he watched the enemy formations for several seconds before taking any action. He called Holden, telling him to keep an extra careful guard as we went down. Then he curved gently to port and dived towards the Messerschmitts which still flew on straight and steady.

Johnnie Johnson and Geoff West positioned themselves on Bader's port side. I lined up on his starboard. There was an enemy for each of us, and several left over for the rest of the boys, if they could get there in time. As we closed I had to steel myself deliberately to rivet my attention on the enemy plane in front which I had selected as a target. Although I knew that our tails were covered I had the strongest instinctive urge to look round behind me.

We closed fast – a little too fast. With half an eye I watched Bader and the second he opened fire I did the same. At the same instant someone shouted 'Break' and I hauled round. The sky behind was full of 109s, guns winking. Where they had come from, God knows. But in another second or two they would have nailed us all. Our own two rear sections were engaged and Ken had brought some of his planes down to join the *mêlée*. The dogfight was hot and furious. Several times I fired my guns; several times I was under hard pressure of attack. No time to worry about results, no time for anything except taut, insistent, concentrated effort to avoid getting in front of an enemy's guns. The penalty for getting caught in that game of tag was death and destruction. And everyone, sooner or later, got caught.

But that day I won through again. The pressure slackened and I was able to ease off and take stock. I called Bader, asking for his position. No answer. Again and again I called, receiving no reply. But my own predicament was still uppermost in my mind as I dived steeply, throttle wide open, towards the Channel coast. Slicing through a gap in the broken cloud, I emerged over the green and grey, light and shade-dappled sea. Sweat-soaked and exhausted I throttled back and set course for Dungeness and home.

Only then did the implications of Bader's silence strike home. Faintly, I heard Woodhall's deep voice calling Dogsbody. I was too low to answer, but Johnnie replied. He told Woodhall that we had had a stiff fight and that he had last seen Bader on the tail of a 109. We straggled back to Westhampnett in ones and twos. Woodhall was waiting at our dispersal point. He still had no news of Bader. Buck Casson was also missing. He and I had joined the squadron together at Doncaster. We had been the last two left of the old pre-war team. Now it seemed that he, too, was gone.

Woodhall questioned me about the fight. I had to tell him that I had seen nothing of Bader from the instant when he opened fire. I told him how, at that precise moment, we were jumped from behind and broke into a fierce and general dogfight. That had been the moment of maximum danger – the moment when a pilot, concentrating all his attention on shooting straight at the enemy ahead, was most likely to fall to the enemy behind. Woodhall went into the dispersal hut to telephone his sector operations room. I stayed outside to smoke a cigarette alone. The thought of Douglas Bader dead was utterly shattering to me. It seemed somehow indecent to imagine his vibrant body and spirit mangled and extinguished in the tangle of metal and flesh which is what remains of a plane destroyed in combat.

Buck Casson's loss affected me in a more conventional way, but it affected me very deeply, as we had been together so long. And it was a terrifying as well as a desolate thing to realize that now I was the last left of the founder group; for it followed, inevitably, that I must also be the next to go. I retreated from these thoughts, taking refuge in the decision that there was at least a chance that both men had baled out over the Channel and were sitting in their dinghies waiting to be picked up. Of course there had been no radio message from either, but I could account for that by supposing that their radio sets had been put out of action in the fight.

These considerations called for action and I immediately asked permission from Burton and Woodhall to organize and lead a search. This I was allowed to do. I asked Johnnie to arrange the fastest possible turn-round for our Spitfires and went into the hut to study the map and work out a search line.

Johnnie came in to tell me that the aircraft would be ready in a few minutes. I explained my plan to him, pointing to the map. If either or both men had emerged from the fight in crippled planes they would make for the coast. It was unlikely that they would have crossed out east of Gravelines or west of le Touquet. We would search just out to sea between those points and most thoroughly of all along the centre of that line, around Gris Nez and Calais. We would concentrate close in to shore, in order to cover first the area of sea from which dinghies might be seen and recovered by the Germans, resulting in the capture of their occupants.

Heppell and West watched and listened, determined to join the search – and I wanted them to come. The four of us were close, tried and trusting comrades. The sortie called for men who would search with the utmost concentration and who would fight with skill and determination if necessary, for we had to count on the probability that the Germans would resent our prolonged presence on their doorstep and send out a superior formation to teach us better manners.

This time we crossed the sea at wave-top level. We made a good landfall and I turned to port when I saw the white line of the little waves breaking on the sands of le Touquet just ahead. Straightening out, I lifted my Spitfire to two hundred feet and set course up the coast, between a mile and two miles out to sea. The others spread out at intervals of about five hundred yards on my left and we began our search. Our progress was followed by the shore batteries and from time to time the black puffs of exploding shells dotted the sky around us. But there was no great danger of a hit from flak fired at that angle. Near Gris Nez a small coastal convoy loomed up out of the haze ahead. There was nothing for it but to keep going and hope for the best. The only thing we could do to minimize the danger of the inevitable flak barrage was to get down as close to the water as possible. Tracer snaked around us as the gunners on the escort ships snarled malevolently at our passing.

We searched diligently until our petrol began to run low. A dinghy was sighted, but on close investigation it turned out to be empty. We also saw a big German rescue float but there was no sign of life near that either. At one point Hepple peeled off and fired his guns at something in the water. I asked what he was doing and he made the surprising reply that he had attacked what looked like a small submarine. At any other moment this intriguing mystery would irresistibly have led to closer investigation. But a submarine search was not on the agenda and I ordered him to rejoin.

When our fuel supply was getting dangerously low I swung north and we flew slowly across the Channel to land at Hawkinge. There we were told that there was still no news of either of the missing pilots. I asked that our planes should be refuelled as quickly as possible and sat down with Johnnie and Geoff West to smoke a cigarette and discuss our next patrol. Hepple stimulated the excitement of an earnest-looking Intelligence Officer with his story about a submarine. In a few minutes our planes were ready and I told Operations that we were going out again, this time to search a little further from the enemy coast. We took off before anyone could interfere with our plans. But soon after we crossed the coast I received a message relayed from Woodhall ordering our immediate return to base. Woodhall had taken a chance in letting us carry out our first patrol, but he knew that there was no real hope of finding anything and he was not prepared to risk a second sortie, when the enemy might have been waiting for us.

Drearily we flew back along the familiar coast line which we had travelled so often behind Bader. Over Dungeness and Eastbourne, where Bader – in defiance of all regulations – used to unclip his oxygen mask, stuff a ready-filled pipe into his mouth and light it, while I pointedly drifted off to a distance which I judged immune from the explosion which might follow. Over Beachy Head and Brighton, over Arundel Castle and the figure-of-eight race course at Fontwell and into the Tangmere circuit. I knew when I had taxied in and stopped my engine that the news was still bad. There was an atmosphere of quiet desolation among the airmen and officers who waited. I was very tired and it was a relief to be told the squadron was released for the rest of the day. Johnnie kindly offered to stay down at dispersal to ensure that everything necessary was done to prepare a full complement of planes for the following day. I drove back to Shopwycke alone and utterly dejected.

I knew that there was one more sortie which I had to make that day. I had to go and see Thelma Bader and her sister, Jill. I had to go because I loved them both and because I felt, illogically no doubt, that I owed them an explanation.

They knew that I always flew with Bader, that whatever had happened we had hardly ever been separated. Now he was missing and I did not have the faintest idea what had happened to him. It seemed to me that, quite apart from my great affection for them, I had to tell them personally that I knew nothing, though I should have done.

I took with me a bottle of sherry and a bunch of flowers and drove slowly through the lanes to their house. The sherry made poor Thelma sick, and I am not surprised. Jill went and cared for her. I left the house and drove back the way I had come and cried as I went. At Shopwycke I found Johnnie and we ordered a bottle of brandy, which we drank.

The Germans were quick to inform the British authorities that Bader, who as it turned out, had parachuted from his severely damaged plane, was a prisoner.

It is almost certain that he would have crashed into the ground with what was left of his Spitfire if he had not had artificial legs. When he first tried to bale out he was held back, half-in half-out, by his right leg, which was trapped in the cockpit. For several seconds he was captive, battered and pummelled by the fierce force of the slipstream as his plane plunged downwards. Then something snapped and he fell free. The something which snapped was the leather harness attaching his right leg to his body. He floated down with one trouser leg flapping around his face. And so, when, at the end of his descent, the ground came rushing up to meet him, he crashed down on one artificial leg and one stump. It is not a pleasant thing to think about, having regard to the fact that a landing on the end of a pilot's parachute was equivalent to a free jump from a height of about fourteen feet. But Bader's tough frame survived the impact.

The story of his subsequent encounter with a German doctor in the hospital at St Omer is grimly hilarious. The doctor first stared in consternation at the empty right trouser leg, but his surprise was tinged with relief when he discovered that the loss of that leg was an old injury.

'Now,' he said, 'we look at the other leg.' It was a moment to remember. Bader raised his buttocks to help the doctor remove his trousers, savouring the glorious anticipation of the shock which the poor man was about to suffer. He was not disappointed.

A few days later, back at Tangmere, Johnnie, Denis Crowley-Milling and I had gone to visit Thelma, still suffering the silent choked-up paralysis of shock. The telephone rang. It was Woodhall, giving the news that Douglas was a prisoner – though, in fact, he was at that very moment in the midst of his first escaping excursion. The sense of relief and release was enormous. The indecent image of Bader broken and dead was removed from my mind and it became possible to look forward again. I was given a week's leave and spent the first two or three days helping Thelma and Jill collect themselves and their belongings for the move to their parents' home at Ascot.

Then I went north to Yorkshire. Cousin Billy Fitzwilliam kindly invited me to join the grouse shoots arranged for that week on his moors near Sheffield and I relaxed and unwound a little, warmly wrapped for a few days in the safe familiar surroundings of my boyhood, before returning south to Tangmere and to shooting of a different kind.

In 11 Group the pace had slackened a bit. We could not know it at the time, but in high places a long hard look was being given at the results achieved by the offensive operations carried out over northern France.

The purpose of these operations, jointly agreed by Fighter and Bomber Commands and approved by Air Ministry, had been expressed as follows:

> The object of these attacks is to force the enemy to give battle under conditions tactically favourable to our fighters. In order to compel him to do so, the bombers must cause sufficient damage to make it impossible for him to ignore them and refuse to fight on our terms.

After the German assault against Russia had been launched there was added to this general intention the specific objective of forcing the Luftwaffe to retain a high fighter strength in the west. But, in spite of all our efforts, this objective had not been satisfactorily achieved. Quite the reverse in fact, British intelligence showed that at the beginning of the Russian campaign there were about three hundred German fighters of the front line in the west. At the end of July fighter units representing a hundred planes had been withdrawn and transferred to the Russian front.

In the six weeks between mid-June and the end of July – the period of maximum activity by our squadrons – Fighter Command lost one hundred and twenty-three pilots and planes. Against this, we claimed the destruction of three hundred and twenty-two German planes. These figures were not compatible with the withdrawal of the Luftwaffe units from west to east. It had to be concluded that our claims were exaggerated – and considerably so. This conclusion was subsequently confirmed when German records became available.

The actual results speak for themselves. Between the middle of June and the beginning of September, Fighter Command lost a hundred and ninety-four pilots and planes in these cross-Channel operations. In the same period German losses of fighter planes based in France and the Low Countries amounted to a hundred and twenty-eight. And of course the effective loss of German pilots was less than that, because they were fighting over territory which they controlled and occupied, so that those who parachuted or force-landed could often return immediately to squadron duty.

However, the RAF commanders of the day had to make their decisions against the background of information they were receiving through the squadron combat reports about damage inflicted on the enemy. Even though this information suggested that we were shooting down two or three enemy planes for every one of our own which was lost, an argument developed in high places about whether the sweeps should continue. In particular, doubts were expressed about whether these daylight raids constituted a sensible and effective use of the RAF's heavy bombers. With hindsight, it is fairly obvious that they did not. The actual effect of the bombing was of negligible significance, the presence of the bombers being primarily intended to bring the enemy fighters to action. Not surprisingly, Bomber Command leaders were less than enthusiastic about having their precious four-engined Stirlings used as bait. In the end a compromise was reached. The heavy bombers would be withdrawn from daylight operations, but the sweeps would continue, the 'bait' being provided by twin-engined Blenheim light bombers, which had no part to play in the mounting nighttime assault on Germany.

All of this, of course, was quite unknown to me or any of the other pilots when I returned to Tangmere from leave on 21 August. I only knew that I had to face up to more fighting and I knew in my heart that I had little enthusiasm for the prospect. It did not occur to me to ask for a rest. Bader's influence had taught me that this was not an acceptable course. Indeed I felt more strongly than ever that I must stick with the squadron, continuing to fight when necessary and helping to pass on to the new pilots the experience and knowledge I had gained, as well as the spirit of aggression with which Douglas had imbued us.

At the same time I subconsciously shrank from battle. The instinct for survival, the inner urge to rest on my laurels, were very strong. I know there were a couple of occasions during the weeks which followed when I shirked the clash of combat at the critical moment. Looking back on it later, I recognized that this was a time of extreme danger for me and also to some extent for the men I was leading. It was the stage of fatigue when many experienced fighter pilots have fallen as a result of misjudgement or a momentary holding back from combat.

In the evenings I turned my attention consistently to the brandy bottle. Night after night, Johnnie and I sat up at Shopwycke and quietly lowered the best part of a bottle between us. Johnnie's iron frame was capable of absorbing almost anything in almost any quantity, though he was inclined to go to sleep in the car on the way back to our billet at Oving and often went on sleeping in the car until the small hours. But when the dawn came he would shake himself and shout for tea and stamp about a bit, making horrendous gargling noises, and be fit for the worst the day had to offer.

Johnnie was in many ways the most outstanding character I served with throughout the war. I think he was greatly affected, as I was, by his early contact with Bader. But he did not really need the steel which Bader could put into a man, because it was there already. He was utterly indomitable and gloriously brave. By diligence as well as by determination he came to be the best fighter pilot in the whole RAF, both as an individualist and as a leader. Yet he never became a desiccated killer-machine, which was the way with some other outstanding fighter leaders. He was always warmly human and his emotions were generous and earthy. His courage was never based on lack of imagination, indeed, his appreciation of the unpleasant things which could happen was often expressed with ribald power and clarity. He was one of the very few who considered it an absolute duty to go on fighting without respite to the end and the only rest he had was forced upon him.

Of course at that time, in September 1941, he was quite unknown, except to our own closed circle. But he was my closest friend and constant companion and he was perfectly delighted to sit up late and help me along with the brandy, if that was what I wanted. This activity, however, eventually came to the notice of Woodhall, who gave me a fatherly talk on the dangers deriving from excessive indulgence of that particular spirit. I cannot remember exactly what he said, but I know that it must have been good, because for a long time I was almost afraid of the stuff. Woodhall had decided to do more than put me off brandy. He had decided to put me off active operations. I was posted as a flight commander to No. 59 Operational Training Unit at Crosby-on-Eden, near Carlisle. When it came to the point my inner relief at being removed from the danger of the 109s was quite swamped by the sadness of leaving 616 Squadron. There was a tumultuous farewell party and the next morning I was poured into the back cockpit of the squadron Magister and flown in a semi-moribund condition to Cumberland.

When we arrived, at about one o'clock, I suggested to Pilot officer 'Huck' Murray, who had flown me up, that we should go to the mess together for a drink and lunch. But the atmosphere of the place put Huck off his appetite. I suppose that we were inordinately scruffy and bleary-eyed. Anyway, Huck did not like the looks we were getting. He swallowed his beer and expressed a desire to return forthwith to Tangmere. And this he did, leaving me utterly bereft. I did not even have my golden retriever, Robin, to console me. He, feeling no doubt just as miserable as his master, was on his way north in the guard's van of a train.

CHAPTER TEN

Intermission

In retrospect I can see clearly how unenviable and difficult was the task of Air Vice Marshal Vincent, who commanded No. 81 Group, with responsibility for all Fighter Command operational training units.

His job was supremely important. He had to ensure that a steady flow of competent pilots was continually available to the squadrons. To provide the necessary training Vincent had a collection of instructors of whom some were so operationally fatigued that they had no enthusiasm for the job, while others preferred the comparative safety of instructing to the hazards of squadron life and were therefore unlikely to imbue their pupils with a proper sense of dash and aggression. Perhaps it was partly in order to counter that situation that Vincent imposed a regime of strict discipline throughout his command. The easy-going habits of dress and behaviour common on all fighter stations were severely discouraged by his unit commanders and Vincent made regular tours of inspection to see for himself that proper standards were being maintained. The imposition of discipline came easily to him, for he had received his early service training in the Marine Corps.

I am afraid that I did nothing to ease the situation for the station commander at Crosby-on-Eden. For one thing, I objected strongly to the strict rules which prevailed on the station in regard to dress, particularly as our quarters were in austere little huts separated from the officers' mess by long stretches of muddy paths. And Robin refused absolutely to recognize the rule which forbade dogs entry to the ante-room or dining room. He had been brought up as a squadron dog and since I had every intention of getting him back to the squadron without delay I refrained from putting any new-fangled ideas of right and wrong into his head.

I found a ready companion in crime at Crosby in the shape of an old friend and fellow Auxiliary, Peter Dunning-White. Having been a pre-war member of London's 601 Squadron he had all the right ideas, still referring to regular officers as 'coloured troops'. He had already been at Crosby for two or three months and his one objective was to get back to a squadron.

Peter and I commanded 'B' and 'A' Flights respectively and the pupils of these flights were all Free French – gallant men who had risked all in escaping from their occupied country for the purpose of carrying on the fight. Many of them were experienced pilots, though naturally they had been out of flying practice when they arrived in England. Their approach to the business of fighter training was uniformly dashing and zestful and this élan combined with lack of practice produced some hair-raising episodes.

A visit of inspection by General de Gaulle was scheduled for 16 October, two-and-a-half weeks after my arrival at Crosby. Peter and I worked hard to train our pupils in preparation for a fly-past. The Frenchmen entered enthusiastically into the spirit of the thing, flying in such close formation that it looked as though their purpose was to scrape the paint off each other's wings. Although frequently frightened to death, I strongly approved of this approach and looked forward with confidence to a stirring performance for the General.

The big day dawned fine but tempestuous, with a strong gusting wind blowing from the north-west. After breakfast Peter and I hurried down to the airfield to ensure that our Hurricanes were all serviceable and lined up in the right order for take-off. I had been warned that Vincent, who was accompanying de Gaulle, felt very strongly about dogs and I put Robin in charge of my batman, with strict orders that he should be rigorously confined.

At the appointed hour Peter and I assembled our pilots in two rows on the taxi-track and placed ourselves at their head. Unfortunately I was unable at the last minute to find my hat. However, this did not seem to matter too much, since we were on the taxi-track not the parade ground and our purpose was to fly not to drill. Peter sportingly agreed, in the interests of solidarity, to appear hatless also. Since neither of us had had our hair cut for rather a long time, and since the gale-force wind was blowing from directly behind us, there is no doubt that we were looking a little dishevelled by the time the General, the Air Vice Marshal and their entourage appeared on the scene. In fact I found myself peering at them through a tangled mass of hair feeling that I must have looked more like a Yorkshire terrier than a Yorkshire Auxiliary Air Force officer.

This hirsute curtain was not so dense as to hide from me the shocked and malevolent looks which I was getting from Vincent and the station commander. I became suddenly conscious of the fact that my trousers, tucked untidily, together with an assortment of maps, into a sloppy pair of flying boots, were of a different vintage from my tunic, that my tunic, with its scarlet lining exposed by the wind, had seen better days; that the brass 'A's on my lapels, which we were no longer supposed to wear at all, were lop-sided, one being about twice the size of the other; and that my tie, instead of being the regulation black, was rather a light shade of blue. All this was brought home to me without a word being spoken, as Vincent's eyes travelled up and down the seventy-six-and-a-half inches of my disreputable person.

De Gaulle, evidently made of sterner stuff, betrayed no similar signs of dissatisfaction as he politely requested me to introduce him to my pupils. The subsequent procession down the line of pilots was reduced almost to a farce by the sudden appearance of Robin, who had evaded his custodian. You have seen a golden retriever which has been running in mud? Well, Robin had passed through plenty of mud during the course of his gallop to join us. He was plastered with the stuff. He was, also, a demonstratively affectionate dog. Tail thrashing, jowls drooling, mud splashing, he joined in the introductions. Seeing the General shaking hands all round, he jumped up to do likewise. He favoured Vincent with a particularly warm greeting. He left his mark on all and sundry before he was eventually captured and dragged away.

After that, it seemed a good idea to get into the air as quickly as possible and this we proceeded to do. We went through a series of pre-arranged manoeuvres, flying as a squadron, with Peter's formation following mine. Then we split into two flights, to dip in final salute. Feeling that de Gaulle's Gallic nature would be stirred rather by dash than by discretion I brought my flight down-wind at maximum speed and at very low altitude. We fairly skimmed the brass hats in our exuberant passing and I was told afterwards that one or two of the frailer spirits were on their hands and knees. They were hardly upright again when Peter and his flight repeated the treatment. I understand that the only member of the party still retaining his phlegm as they entered their cars and drove away was the General himself.

That Vincent's phlegm had gone up in smoke was made apparent the moment I landed. One of the station commander's administrative minions was waiting for Peter and me on the tarmac. Chattering with rage like a startled magpie he let it be known that in the opinion of his masters we were unfit to live. He seemed to think that if anything I was even less fit to live than Peter, which seemed a little unfair, since his appearance was no less unorthodox than my own. Perhaps it was Robin's part in the proceedings which contributed to the special disfavour in which I found myself.

After the staff officer had left us Peter and I decided that in the course of his tirade a degree of priority had been placed on the need for a hair-cut. Very well, we would attend to that matter right away. There was to be no more flying that morning, so we would go into Carlisle and visit a barber. After shutting Robin up in my room we set course in Peter's nice little 25 h.p. Rolls Royce coupé ('my baby Rolls', he used to call it) to get the job done. It was about half-past twelve when the barber had done his work, so it seemed reasonable to call in at the bar of the station hotel for a glass or two of gin before going back for lunch. This took a little time and we did not reach the mess until a quarter to two. We were immediately set upon again by the administrative minion, who wished to know why we had been absent when de Gaulle and Vincent were brought to the mess for sherry at twelve-thirty. He added that although I had been absent my damned dog had of course been present, having made a precipitous and mud-spattering entry into the ante-room, inciting the rage of the Air Officer Commanding and causing grave discomfiture to the station commander.

The upshot of the morning's work was, surprisingly, entirely favourable to me. Luckily Vincent and the station commander acted in the heat of their rage and chagrin. Vincent proclaimed that he would not tolerate my presence in his command for another moment. I was to be removed. He did not care where to, just so long as I went and went quickly. This entirely welcome information was passed on to me in an interview with the station commander after lunch. I acted with lightning speed. Although the appointed term of duty for an instructor at operational training units was a minimun of six months, I had already started planning my escape route. I had ascertained that Ken Holden, still commanding 610 Squadron, which had been moved north from Tangmere to Leconfield, required a new flight commander and was in favour of giving me the job if I could be extricated from 81 Group.

It was a race against time. I knew that when Vincent calmed down he would realize that by removing me from his Group he was doing me nothing but a favour and I had therefore to get things organized before he countermanded his order. I telephoned Ken Holden, who immediately telephoned the Senior Personnel Staff Officer at 12 Group, who immediately telephoned his opposite number at Fighter Command, who authorized an immediate posting. By the following morning the whole arrangement was officially confirmed by signal and I departed from Crosby-on-Eden at top speed. I had been an instructor for three weeks exactly. So far as I was ever able to discover, this established a record for brevity which remained unbeaten in Fighter Command throughout the war.

Poor Peter Dunning-White! He claimed with justice that he had done every bit as much as I to merit dismissal. Yet he had to stay on, although he had already spent several months in the place. I advised him to get hold of a good big dog – and better luck next time.

CHAPTER ELEVEN

Typhoon Command

My stay with 610 squadron at Leconfield was brief but entirely happy. It was a cheerful station, presided over by a most delightful wing commander called Tim Morrice, who had won a Military Cross in the Royal Flying Corps in the First World War and retained a fine youthfulness of spirit. He used to fly about in a Tiger Moth, which was his pride and joy. One day when he took off over our dispersal hut we decided to give him a fright by shooting off a volley of Verey light cartridges at him. By an unfortunate fluke and greatly to our dismay one of these shots was all too accurate and struck home in the engine cowling. Tim landed in a cloud of red smoke. Luckily he got down all right and no great damage resulted to the plane. We waited in fear and trembling for retribution, but Tim treated the episode as a huge joke. He was the kind of station commander we all approved of.

One of Tim's many enthusiasms was the playing of rough and rowdy games in the mess after dinner. On the night of 18 November he was at his most boisterous in this respect. The beer was flowing freely and everyone was in fine form. Late on, a game of rugger developed. Tim picked a team and I picked the other, a waste-paper basket was used as a ball. We went at it ding-dong, hammer-and-tongs. After several minutes with no score an opening came my way and I scooped up the basket and made for the line. Two or three of the opposing team tackled me, Tim to the fore, and I crashed to the ground at the bottom of the pile. Someone dragged the battered remains of the waste-paper basket from underneath us and the game moved on.

I started to get up, felt a sharp pain in my left leg and sat down again with a bump. Try again. Same result. The game was stopped and everyone crowded round. I was helped to my feet – or rather to my foot, for my left leg would take no weight. The station dental officer, who had been taking part in the game, examined the leg and everyone thought this was a wonderfully funny joke, myself included. He pronounced it broken and appeared rather pained when I asked how the hell he knew – it was my leg, not my mouth which was causing the trouble and would someone please find a doctor. But there was not really much doubt about the dentist's diagnosis, which the doctor confirmed as soon as he had been brought to the scene from his bed. By the time the ambulance arrived I had swallowed two or three whiskies prescribed by my friends and I went off to hospital in Beverley already well anaesthetized.

Next morning when I surfaced things did not seem so funny. A nurse told me very severely that I had been in no condition to have my leg set the previous night. And she asked me, severity now tempered by genuine curiosity, how on earth it had come about that my trousers had been on inside out. I told her that I always wore them that way in the evenings, so that the beer stains would not show in the day time. From that moment she treated me as a dangerous lunatic and we maintained a friendly feud until I was let loose, leg cased in plaster, a few days later.

When I got back to Leconfield, Tim Morrice hinted to me that there was a possibility that I might soon be moved to take command of a squadron.

There was much movement in Fighter Command at that time. Sholto Douglas now had a hundred squadrons at his disposal and there was something of a general post within the command as a result of the formation of 'national' squadrons – Canadian, Australian, New Zealand, Polish, Czech, Belgian, Norwegian and French. There were even three American squadrons in the command, with aircrews made up entirely of United States volunteers who had made Britain's fight their own. Many of us thought it a pity that so many of these national squadrons had been formed – there were thirty-four of them by December 1941 – since we had experienced the fine sense of comradeship engendered in the mixed squadrons in which, or with which, we had served. But it was a necessary political move and fortunately there were still plenty of fine men left over from the Empire and occupied countries to serve in RAF squadrons.

It was not only the formation of these national units which resulted in many new appointments at flight commander and squadron commander level. There was also an accelerating movement of pilots from Fighter Command to the Middle East. Indeed, some historians have taken the view that this movement was too long delayed, criticizing the extent to which the UK fighter force was maintained at a time when modern planes and experienced pilots were desperately needed in the Western Desert and in Malta. To this was to be added, after the entry of Japan into the war that December, a pressing need for pilots and planes in the Far East.

Despite all that, it was a shock to me to be told that I was being considered for command of a squadron. I had started out the year as a mere pilot officer – the lowest form of human life. I was barely five months past my twenty-first birthday and I knew of no one with the rank of squadron leader at that time who was less than twenty-four or twenty-five years old, and even that was still considered young. No wonder I faced the prospect with some trepidation.

However, this was nothing to the trepidation I felt when Tim Morrice told me I was to go to group headquarters to see the Air Officer Commanding, Air Vice Marshal R.E. Saul. 'Birdie' Saul had a reputation as something of a holy terror. A handsome, florid-faced man with fine, prematurely-white hair, his

The author, taken at Kirton-in-Lindsay at the end of 1941

This page
Top: Doncaster,
September 1939 – a
bouquet of dandelions
from Sgt Dale after my
first flight in a Gauntlet

Right: September 1939
– 24 hours leave
My father and mother

Opposite page
LECONFIELD,
FEBRUARY/MARCH 1940
Dennis Gillam (*top
left*)
George Moberley
(*top right*)
Roy Marples
(*bottom left*)
Teddy St Aubyn
(*bottom right*)

Right: Spitfires taking off for convoy patrol, early 1940

Below: Dunkirk – the German target

Opposite page
Top: Leconfield, August 1940 – Tom Murray and Buck Casson 'at readiness'

Bottom: Ground crew – Corporal Durham top right

19 AUGUST 1940   POST PRANDIAL DEPARTURE FOR KENLEY
Top: Jack Bell, Teddy St Aubyn, Bill Walker
Above left: Dennis Gillam
Above right: Self, George Moberley, Roy Marples

BATTLE OF BRITAIN
Top: End of Heinkel III
Above: Remains of Spitfire

Bus stop

AT FOWLMERE, WITH BADER'S 'BIG WING', SEPTEMBER 1940
Ken Holden (*top left*)
Phil Leckrone, a volunteer from the USA (*top right*)
Colin Macfie (*left*)
Jerry Jones, after successful combat, 27 September 1940 (*above*)

Opposite page
TANGMERE 1941
Diana at Oving (*top left*)
With Bader, walking in
after a sweep (*top right*)
Johnnie Johnson
(*bottom left*)
Air Marshal Sir William
Sholto Douglas, C-in-C
Fighter Command, with
Brian Kingcome, later
my CO in Italy (*bottom
right*)

This page
Top: October 1940 –
with John at Cawthorne

Right: Tangmere 1941 –
with Buck Casson
outside the squadron
house at Oving, plus
Mac and Robin

Opposite page
Top left: January 1942 –
56 Squadron at Duxford.
Michael Ingle-Finch is on my
left
Top right: John Grandy was
Station Commander at Duxford
1942 (here pictured some years
later as an Air Marshal)
Bottom: September 1942 –
second anniversary of the Battle
of Britain. *From left to right*:
Max Aitken, 'Sailor' Malan, Al
Deere, Sir Hugh Dowding,
Elspeth Henderson MM,
Richard Hillary, Johnnie Kent,
Brian Kingcome

This page
Above: Matlaske, autumn 1942 –
'Stimmie' Stimpson, with a
Typhoon of 56 Squadron. He
later commanded 601 Squadron
in my Wing in Italy
Left: Italian bomber-eye
view of Hal Far airfield,
Malta, where 324 Wing
was based for the Sicily
invasion

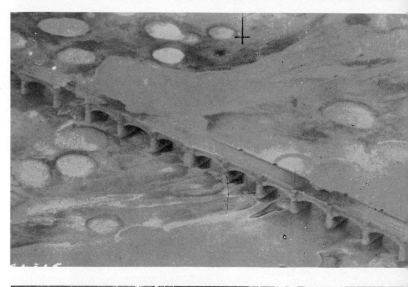

This page
Top: Baltimore light bombers in action near Salerno, September 1943

Middle: Desert Air Force interdiction – railway bridge at Fano, north east Italy, damaged by Kittyhawk fighter bombers

Bottom: Pierced Steel Planking – the stuff of which our runways were made on top of the Italian winter mud

Opposite page
Top: Clare Booth Luce visited 244 Wing at Bellaria just before the final offensive. On right of picture, Johnnie Gasson, CO 92 Squadron

Bottom: Douglas Bader, recently returned from prison camp, climbing into his Spitfire before the victory fly past in 1945

Group Captain Harry Broadhurst, DSO (and bar), DFC, AFC
(Portrait by Captain Cuthbert Orde, 1941)

presence was imposing and he was not in the habit of doing anything in his approach to junior officers to ease the awe in which he well knew he was held. I set off to see him with my heart in my plaster cast.

His opening remarks, after I had stumped in and saluted, were not reassuring.

'Broke your leg in the officers' mess eh? Damn silly thing to do!'

'Yes, sir.'

'Still got the plaster on, have you?'

'Yes, sir.'

'Well, what the devil have you been doing flying a Spitfire in that condition. Doctor say you could?'

Oh God! How on earth did the old devil know about that?

'Well, no sir. But I found I could manage quite well, so I thought I might as well fly.'

'Quite right. But don't do it again until that plaster's off – particularly now I know about it. Sit down.'

He told me he had selected me to command the first Spitfire squadron to go overseas: No. 601 Squadron. It was going to Malta, where the fighting was intense. The squadron would be embarked with its planes on an aircraft carrier and would fly off when about three hundred miles from the island. Hurricanes had done it, so Spitfires could too. What did I think about that?

My heart did a loop. Malta . . . first Spitfire squadron overseas . . . aircraft carrier. Blimey!

'I think it's marvellous, sir.'

'Well I can't send a bloody cripple on a job like that. Teach you to break you leg playing damn-fool games in the mess.'

Oh God! What a hell of a let-down.

'Never mind. I've got a consolation prize for you. You can take over 56 Squadron. That's at Duxford, re-equipping with Typhoons; first squadron to get them. Get down there as soon as you can. You're to take over from Hanks. He's going to stay at Duxford as wing commander. And don't fly until your leg's out of plaster. Goodbye and good luck.'

Goodbye and good luck indeed. Typhoons were said to be the fastest things flying and 56 Squadron, one of the most famous in the Royal Air Force! Look out Duxford, here I come! I took the AOC at his word, said goodbye to 610 Squadron and to Leconfield and journeyed down to Cambridgeshire as quickly as I could.

'Prosser' Hanks took me into the ante-room at Duxford and introduced me to the officers one by one. Then he and I stood by the fireplace with our drinks and talked to the flight commanders, Peter Gifkins and Michael Ingle-Finch. I was ill-at-ease as I saw the junior officers, standing around in groups with their beer tankards, constantly glancing at me out of the corners of their eyes. I recalled the merciless secret inspection to which I and my friends had in the past subjected new commanding officers. And I realized that I was now at the receiving end of similar treatment.

I pressed the bell and when the steward came I ordered pints of beer all round. Then, with a determined effort, I left the security of my little circle with Hanks and the flight commanders and walked out to get to know my pilots. They were a more than averagely mature and altogether a self-confident lot. I discovered later that I was at that time the youngest officer in the squadron, but they had probably made this discovery already and no doubt they were looking for any signs of weakness or lack of authority in this juvenile fellow who had been foisted upon them.

I soon discovered that enthusiasm for the Typhoons was lukewarm, to say the least. The first of these new planes had been delivered in September and since then there had been nothing but trouble. On 1 November there had been a fatal accident, when an officer of the squadron had unaccountably dived straight into the ground in a Typhoon from a height of three thousand feet. It was later discovered that he had been poisoned by carbon monoxide and all Typhoons were immediately grounded. By the time I reached the squadron on 22 December a small number had been modified and passed fit for flying. Meanwhile the squadron had to maintain a state of readiness with its few remaining Hurricanes – a tedious and unrewarding business unlikely to yield any results in the form of contact with the enemy.

It was perhaps not surprising that the pilots should be beginning to wish that their squadron had not been selected as the first to get Typhoons. It appeared, certainly, to be an honour of dubious advantage. Yet the attitude was worrying me, because it was a matter of tremendous importance that these new planes should be successful.

In the Luftwaffe, too, a new single-engined fighter was going into service in units facing Fighter Command over the Channel. It was the Focke-Wolf 190 and it soon became evident that its performance was outstandingly good. Powered by a BMW engine of one thousand eight hundred h.p. it had a top speed of over four hundred m.p.h. It could easily outstrip the best Spitfire in every department and Fighter Command was in for a sticky time until the right answer was produced. In due course that right answer turned out to be the Spitfire Mark IX. But in the winter of 1941/42, when I was taking over 56 Squadron, it seemed that the Typhoon represented our principal chance of achieving parity.

The Typhoon had been designed by Sidney Camm, chief designer at the Hawker Aircraft Company and creator of the immortal Hurricane, as a machine which could take advantage of the two thousand h.p. engines being developed by Rolls Royce and Napier just before the war. Eventually the Napier 'Sabre' engine was chosen. This massive motor was in itself revolutionary. It had twenty-four cylinders, set in an 'H' section, and sleeve valves. It had a power output in excess of two thousand h.p. Driving a four-bladed airscrew fourteen feet in diameter, this engine propelled the Typhoon through the sky at a top speed of well over four hundred m.p.h. At fairly low altitude the plane cruised along comfortably at an indicated airspeed of between two hundred and eighty and three hundred m.p.h., compared with about two hundred and thirty m.p.h. which was the equivalently comfortable indicated cruising speed of the Spitfire Vb. The Typhoon was armed with four 20mm cannons, mounted in the wings.

Compared with the Spitfire, the Typhoon was a massive, almost brutal-looking machine. While I could stand on the ground beside a Spitfire and lay my hand on the top of the cockpit canopy, the cockpit of a Typhoon towered far above me. Instead of one step onto the wing and a second step into the cockpit of a Spitfire you had to carry out a minor mountaineering feat to climb up into a Typhoon, eventually entering the cockpit through a door, like that of a motor car.

That was the warplane the operational development of which now became my very personal responsibilty. As I talked to my pilots that first evening in the mess at Duxford I noticed that there was no disposition on their part to make the job sound easy. But, then, that had always been the way with pilots talking about a new type of plane to the uninitiated. It was normal to lay the difficulties and hazards on a bit thick. So I was not unduly impressed, or depressed either.

The Typhoon aircraft dominated my life throughout 1942. The objective of putting it on the map as a successful fighting machine took on, for me, the nature of a personal crusade. In this I was supported over the months by an exceedingly enthusiastic and squadron-proud group of pilots. In 56 Squadron that year there was no odd man out. If one cropped up, then he went out. Whatever we did, on the ground as well as in the air, we did as one. When we went out in the evenings, we went as a squadron, each putting the same amount of money in the pool, which was then handed to a 'treasurer' appointed for the occasion. We were tremendously proud of the squadron, with its magnificent tradition – it had been one of the very best in the First World War, including among its pilots at that time such legendary aces as Ball and McCudden. And we insisted on being proud also of our Typhoons, despite the unending difficulties, disappointments and, sometimes, dangers with which they confronted us.

Having at last disposed of my plaster cast, I made my first flight in a Typhoon on 2 January 1942. Taxiing out, I felt as though I was about to try to take off in a steam roller. But once off the ground I soon responded to the sense of excitement engendered by the plane's speed and power. I did a couple of rolls over the airfield and came in for a careful landing.

My pilots crowded round to get my first reactions. What did I think of her? 'I think she's great,' I said. 'The speed is really exciting. And what a steady gun platform she must make. But one thing's certain. We're not going to war until the rear view is improved. That will have to be put right.'

The sloping cover behind the cockpit on those early Typhoons was made of material which was solid and opaque not, as on the Spitfire and Hurricane, of transparent perspex. To make matters worse a slab of equally opaque metal armour plating extended almost from one side of the aircraft to the other immediately behind the pilot's head. The result of all this was that, however far you leant over to peer back, there was a blank spot behind you which extended through a cone of about 160 degrees. I immediately made up my mind that this was utterly unacceptable. A fighter pilot who could not watch his tail was a dead duck, no matter how fast his plane.

I put in an official report, stating this view very strongly. It was a fortunate thing for me that at about this time Duxford station was taken over by John Grandy, then, I believe, the youngest group captain in the RAF. He was an experienced fighter pilot and he had an original mind and was not in the least afraid of pressing his views on higher authority. We saw eye-to-eye on most things and he gave me wonderful support and leadership in the difficult times ahead. He started off by agreeing about the inadequacy of the rear view.

I think my squadron pilots were impressed by the stand I took in the matter and that this was the beginning of the mutual trust which developed between us. However, there were others in more powerful positions who undoubtedly took the view that I was nothing but a damned nuisance – and a young, jumped-up damned nuisance, at that. Already the production of Typhoons had been seriously delayed by the need to modify the aircraft for the purpose of avoiding carbon monoxide poisoning. If now a major reconstruction was to be undertaken to improve the rear view, further and longer delays were inevitable. In any case, there was enough trouble with the engine, without messing about with the airframe. That was the response which arose from some quarters when my report went forward.

On 10 February a big conference was held at Duxford to thrash out this and other matters. Air Vice Marshal Ralph Sorley, who was the officer at Air Ministry responsible for technical development, presided. Sidney Camm was there and so were an assortment of senior representatives and staff officers from Napiers, the engine manufacturers, the Gloster Aircraft Company, where the Typhoons were being built, Fighter Command and 12 Group. There I sat, the youngest squadron commander in the business, confronted by all that majesty and power. Not surprisingly I was shivering in my shoes.

However, I was determined not to give way and when I was called upon to give my views I stated simply that if I had been asked to go to war in 1940 and 1941 in an aircraft with similarly restricted rear view I should have been dead long ago. I said that I did not believe that any experienced fighter pilot would disagree with me in stating that the matter must be put right before we were made operational. I remember Camm muttering something to the effect that 'the aeroplane's so fast you don't have to see behind you.' But there was much sympathy for my views among the officers present and there was no real opposition from the manufacturers once they were convinced that the modification was really necessary for the safety of pilots. Quite the contrary: Camm's staff redesigned the rear end of the cockpit in double-quick time, providing us with two slabs of armoured glass behind and above each shoulder and a transparent perspex cover behind that.

Round one to the squadron.

Throughout February and March we flew hard with whatever serviceable aircraft we could muster. Our remaining Hurricanes had been taken away from us and the squadron had been made non-operational. Our sole purpose was to flog the bugs out of the Typhoons as quickly and thoroughly as possible.

There was a very high standard of pilotage in the squadron, which was fortunate, because engine failures and other troubles provided plenty of scope for serious accidents. I became very attached to Mike Ingle-Finch, the 'B' Flight commander. He was an excellent pilot and an amusing companion and I put great trust in him. When, towards the end of March, I succeeded in getting a brilliant former sergeant pilot, Brian Fokes DFM, to command 'A' Flight I had two outstanding men in positions of authority beneath me. Of the other pilots, two Canadians stood out – Bob Duego and Wallie Coombes. Another memorable character from across the Atlantic was Ken Macdonald, who had been brought up in Brazil; he was to end up commanding a squadron in my Wing in Italy and was killed at Salerno. But there was hardly a pilot in the whole squadron whom I would not have classified as above average.

On the administrative side, too, I was wonderfully well served. The adjutant was Basil Hudson, a City of London business man who steered me out of trouble on many stormy occasions. The intelligence officer was 'Kay' Somerville, who in civilian life ran the 'K' Shoe Company, of which he and his family were I believe, principal shareholders. Finally, there was Harold Wareham, the engineer officer. He was nearly driven out of his mind by the difficulties he faced, particularly with the Sabre engines, and he felt strongly his responsibility for our safety. His was a notable contribution to the eventual success of the Typhoon aircraft.

In February Air Marshal Sir William Sholto Douglas, Fighter Command's C-in-C, visited us. From that day onwards I felt the greatest admiration and affection for his massive and seemingly pugnacious presence, which, loaded down with all the braid of an Air Marshal, illuminated by the decorations of a First World War fighter ace, might have combined to inspire more alarm than affection. But one of the most notable of his many qualities was to make you understand and appreciate that he really cared deeply about what you were doing. He showed an intense interest in the progress we were making, in the difficulties we were facing and in our opinions about the Typhoon's future. He made me feel that my views really mattered to him – as, indeed, I suppose they should have done, though that was not always the way. When he left us my own morale and that of the whole squadron was up several points.

During March the re-equipment of two other squadrons with Typhoons began. One was 266 Squadron, a newly-formed Rhodesian unit, commanded by Charles Greene and including among its officers Pilot Officer Ian Smith, later to be Prime Minister of his country. The other was 609, my brother John's old squadron. The latter was commanded by Paul Richey, who was not only a fine pilot but had also an almost unrivalled capacity – and reputation – for burning the candle at both ends. He had an additional claim to fame, the authorship of the first, and extremely successful, book about air fighting in World War II. Its simple title was *Fighter Pilot* and it had been written during the summer of 1940 while Paul was recovering from wounds received towards the end of the blitzkrieg in France, where he had flown Hurricanes with 111 Squadron. He was, incidentally, married to the sister of John's CO in 609 Squadron, Michael Robinson.

In order to make room for these two units at Duxford, 56 Squadron was moved on 30 March to a satellite field called Snailwell, just north of Newmarket and some eighteen miles from the parent station. Although the amenities there were primitive, the move suited us well, for we enjoyed being on our own.

In April, Denis Gillam, my old flight commander from the early days in 616 Squadron, was posted to Duxford as Wing Commander Flying and we began to train as a Wing, when serviceability allowed. Gillam had earned a reputation for quite outstanding gallantry since I had last served with him. He had been in charge of a squadron at Manston engaged in daylight attacks against enemy shipping. It was an excessively dangerous job and losses were high. But by a miracle Gillam, who had, as was always his way, led from the front, survived. In the end he was quite badly wounded, but managed to get back to base. No doubt that wound saved his life, for had he continued at that particular game much longer he must surely have been killed. With his arrival and the beginning of Wing training it seemed as though we might at last be getting near to operations. Unfortunately it was at this time that more serious trouble developed.

One of the constant sources of difficulty with the Typhoons had always been the matter of oil cooling. This made life particularly complicated, because it involved both the aircraft and the engine manufacturers, the design and positioning on the airframe of the air inlet for cooling the oil being an important factor.

The squadrons were asked to carry out a series of full throttle climbs to maximum altitude, reading temperatures carefully in the process. A pilot's natural inclination after completing his climb was to come down again as quickly as possible. It was in the course of such descents that two or three aircraft disintegrated. The pilots, of course, were killed, so there was no explanation to be had from them. But the wreckage showed one thing clearly: disintegration had in each case been caused by the breaking off of the tail unit. This development, as you may imagine, was not good for morale. A swarm of technicians descended upon us from Hawkers and Glosters. Our aircraft were grounded for minute examination of the joints where the tail units fitted onto the fuselages, but no signs of weakness were found.

Philip Lucas, chief test pilot for Hawkers, and Jerry Sayer, his opposite number from Glosters, came to stay with us. With great courage they demonstratively did everything which anyone could think of to make an aircraft fall apart and they did it within full view. One thing was discovered. The Typhoon, with its great weight and power, could soon reach 'compressibility' in a vertical dive. This was the first we had heard of compressibility, but it turned out to be the phenomenon which later came to be associated with the speed of sound, or 'sound barrier'. Different aircraft reach this critical point at different speeds, according to the design and shape of their wings. In the Typhoon it had the effect, when in a steep dive, of making the nose go down and beyond the vertical. And no action taken by the pilot would rectify that situation until the speed had decreased.

The theory therefore arose that the dead pilots, having found themselves in this unfamiliar and frightening situation, had automatically throttled right back and at the same time wound back hard on their trimming gear in an effort to bring the nose back up. Thus when their aircraft slowed down sufficiently to allow the control to function normally, they were trimmed so tail-heavy that the aircraft had whipped up from dive to climb with excessive violence, imposing breaking strains on the structure.

This theory at least gave us something positive to cling to. The best thing seemed to be to avoid going too fast in a dive. This all Typhoon pilots with a properly developed sense of survival did in future. Long after the war, however, Bill Humble, a Hawker test pilot who spent weeks at that time gallantly, but unavailingly, trying to tear Typhoons to pieces, told me that he was convinced that the trouble was 'elevator flutter'. That is a condition which from time to time arose in certain types of plane and for no reason known to man so far as I can discover. And if you got elevator flutter badly enough you had about a second of your life left to you, so there was hardly time to send a message back to base about it.

While we in the Duxford sector were contending with these disagreeable mysteries the Spitfire squadrons on the south coast were trying, without too much success, to counter a new type of German activity. This took the form of low-level attacks by Messerschmitts and Focke-Wulfs against coastal targets. The attacks had more nuisance value than anything else, for the German pilots, streaking in from the sea, dropped their bombs or fired their guns on the first target which offered, without worrying too much about its military value. But there was loss of life and property to an extent which demanded effective counter-measures. And Fighter Command was on the spot, because the Spitfire Vbs with which the squadrons were still equipped, were too slow to catch the latest German fighters at low level.

The situation was seen to offer an opportunity for our Typhoons. We had the speed to overtake anything the Germans could send over – and the lower they flew, the greater our advantage. Thus in the last days of May I was ordered to attain maximum serviceability in preparation for an early detachment to a south coast station. Great was our excitement and jubilation. At last we were to have the opportunity to show what our planes could do. When the operational movement order arrived on 29 May it said that the squadron would move in two flights the next day – one to Tangmere and the other to Manston.

Naturally I decided to go to Tangmere myself. Mike Ingle-Finch, un-fortunately, was out of action after coming into contact with an immovable object with his car while driving home from a party, so I took 'B' Flight. Denis Gillam, sniffing the scent of battle, took 'A' Flight to Manston, his own old base and stamping ground. Within an hour or two of landing at West-hampnett I had my orders from Operations. There were certain peak periods of the day for the hit-and-run raids and during these periods we were to maintain low level standing patrols just off the coast. And so, at 5.30 p.m. on 30 May, I took off with one other aircraft, flown by Pilot Officer Norman Donniger, on the first operational patrol carried out by Typhoon aircraft. How we prayed that it would bear fruit! I well knew that if we had the luck to get in touch with a German raider we could knock it down. But it was not to be. We patrolled up and down the line from Selsey Bill to St Catherine's Point until petrol ran low and returned to base disappointed. On the following day I flew two more such patrols, with the same result.

On 1 June we got action, with tragic results.

The first I knew about it was a telephone call from Operations telling me that two of my 'A' Flight pilots had been shot down over the sea south of Dover. They were Pilot Officer Bob Duego and Sergeant Pilot Stuart-Turner. Both were reported missing. Without waiting for further information I took off and flew flat out to Manston. I was unutterably depressed, not only by the loss of two fine pilots, but also by the thought of the set-back to our hopes of proving the superiority of our Typhoons.

Soon after landing at Manston, my depression was changed to anger. It transpired that Duego and Stuart-Turner had not been shot down by the enemy at all, but by Spitfires. Naturally our pilots had suspected no hostile intent on the part of the Spitfires until the moment they opened fire. And then it was too late for evasive action. However, by good luck, Duego was not killed or totally incapacitated when the cannon shells hit and severely damaged his plane. He succeeded in baling out and was picked up from his dinghy after two hours in the sea. He was taken to Dover hospital, suffering from burns and other wounds, where he was able to give an account of the incident.

His section had been scrambled to intercept a small raid approaching Dover at medium altitude. He flew the courses given him over the radio by the controller, climbing as instructed to seventeen thousand feet. He was ordered to circle at that altitude a few miles south of Dover. Soon he saw two aircraft approaching, which he quickly identified as Spitfires. He resumed his orbit and watched the friendly planes as they curved towards him. The Spitfires joined the circle and closed in. Mistaking the Typhoons for Focke-Wulf 190s, they opened fire simultaneously. Stuart-Turner's plane turned on its back and dived vertically into the sea. It was a heart-breaking incident. But I was quickly satisfied that Gillam, no less infuriated than I, was taking the matter up at group and command with a satisfactory degree of heat and urgency.

I returned to Westhampnett and for a further week we maintained our low-level patrols. But the enemy did not oblige. Perhaps their Intelligence department had warned them of our presence. Anyway, it was decided at Fighter Command that we should return to the Duxford sector to continue preparations for Wing operations.

And so, on 7 June, we landed back at Snailwell. I am glad that I did not know at that first time that our frustrating experiences during those first operational sorties with Typhoons were to be a forerunner of future events.

CHAPTER TWELVE

High Spirits, Low Flying

With three squadrons of Typhoons ready for action and others soon to be formed the problem of how best to use them became the subject of urgent debate at Fighter Command.

There was a strong body of opinion at Command, and at 11 Group also, which favoured keeping them away from the big sweep operations across northern France, now building up again. Influential in this school of thought was Group Captain Harry Broadhurst, the brilliant fighter leader who at that time occupied the key post of Group Captain Operations at Command. He took the view that the Typhoons could not be properly fitted into offensive sweeps. They were not suitable for close escort work and, because their performance fell away very quickly after reaching an optimum at about eighteen thousand feet, they were not suitable for high cover either. The three squadrons were only a spit in the ocean, compared with the seventy-odd squadrons of Spitfires in the command. Confine the Typhoons to special duties and keep them clear of the major operations: that was Broadhurst's recommendation.

Naturally I was opposed to that concept. I insisted that the Typhoons had qualities and capabilities which made them suitable for certain special roles in large-scale offensive operations. I proposed, for instance, that an ideal way of using the Wing would be to send it in for a high-speed sweep round the rear of the main formations as they were withdrawing. We could go in at about twenty-one thousand feet, gradually losing height as we swept around, so that we would be at optimum altitude during the critical stages of our passage. We would attack anything we saw on our way through, not sticking around for any dog fights, for we knew that the Messerschmitts could out-turn us, but taking advantage of our superior speed to pounce and get away. This was a pattern the Germans had used with success when they were flying planes with superiority in speed but inferiority in the turn. Now let us do likewise.

The argument came to a head during a conference at Fighter Command, convened by Sholto Douglas in June. For some reason Gillam could not be present. I went from Duxford with John Grandy and Paul Richey. The conference was a small one, the three of us from Duxford facing Sholto Douglas, Broadhurst and two or three other officers from command headquarters. Broadhurst gave his views and I was then asked, as the person present who had done most flying in Typhoons, to state my opinions.

After I had done so, Sholto Douglas thought the matter over for a few moments before speaking. Then he gave his decision:

'Well, Broadie,' he said, 'you have no experience of these planes, after all. I think we'll see if there's anything in what young Dundas says.'

I could see that Broadhurst was far from pleased by this pronouncement. And I do not blame him. He was an officer of vast experience, clearly and generally recognized as being outstanding as a leader, as a tactician and as a pilot. He was also, as Group Captain Operations, directly responsible for formulating operational policy within Fighter Command. And here he was, forced to give way to the opinions of a jumped-up young auxiliary officer of small experience and no importance. The repercussions from that incident were to affect me severely almost exactly one year later. And looking back on the events of the next three months it seems obvious that Broadhurst had been right and I had been wrong.

We took part in several sweeps during that period, but never succeeded in getting to grips with the enemy. We did, however, on several occasions get to grips with Spitfires. Lieutenant Eric Haabjoern, one of two Norwegian pilots who had been posted to the squadron in June, was shot down by a Spitfire over the Channel on 30 July, while he was limping back home with a failing engine. Fortunately he baled out and was picked up. There were other incidents of a similar kind suffered by 266 and 609 Squadrons.

It was evident that the silhouette of the unfamiliar Typhoon was all too similar to that of the familiar Focke-Wulf 190. And as the Spitfire pilots saw far more 190s than Typhoons they tended to attack first and ask questions later. Nevertheless the situation caused great bitterness. Worse than that, it led the planners at Fighter Command to relegate the Typhoons to more and more insignificant roles, as far removed as possible from the centre of affairs.

On only one occasion during those weeks was 56 Squadron properly engaged during a sweep. That was on the day of the ill-fated Dieppe raid. As usual we were kept well away from the main theatre of activity, sweeping round to the north and east of the battle, ostensibly for the purpose of intercepting German planes making for Dieppe. And for once the plan worked out for us bore fruit. Several bombers were sighted below, Gillam took two squadrons down to the attack, ordering me to stay above and cover him. The resulting action was an inconclusive affair, with no casualties on either side. My job was to keep my squadron intact between Gillam and the Germans above, which kept nibbling at us and then zooming up again. We were on the defensive and if I had made any effort to improve my tactical situation it would have involved leaving the lower squadrons uncovered. We had to go home content with the fact that we had done our job, while Gillam and his boys below shot down several of the enemy bombers. At least the Wing had at last scored.

Soon after the Dieppe affair the Typhoon Wing was broken up. It had finally been decided that our fast and massively sturdy planes were best suited for ground attack role, or alternatively as bomber interceptors. In any case, the Mark IX Spitfires, powered by a development of the Merlin engine with a two-stage supercharger which gave a vastly improved performance at high altitude, were at last going into service. They were undoubtedly superior to the Typhoon for the cut-and-thrust of fighter-versus-fighter warfare high above occupied France, for their enhanced performance had been gained at small sacrifice to the Spitfire's magnificent manoeuvreability.

And so, on 25 August, we moved to Matlaske, a satellite field in the Coltishall sector, a few miles north of Norwich. There we had our own squadron mess in an attractively converted mill-house on Lord Walpole's estate and we were very happy, except for the lack of action. All of us had devoted much effort and enthusiasm into the development of our Typhoons and we were naturally disappointed that we had so little to show for it. This mood was reflected in an increasing tendency among the officers of the squadron to drink and be merry. No doubt I was often the ringleader in the organization of parties. At the same time I was aware of an increasing difficulty in maintaining the right degree of discipline in the squadron.

There were several officers in the Coltishall sector whose personalities added to the convivial atmosphere of the place that autumn. Johnnie Johnson was commanding 610 squadron on the other satellite field. Max Aitken commanded a very successful night fighter squadron at Coltishall. Another auxiliary, Roger Frankland, was Senior Operations Officer. And the Wing Commander Flying was Johnnie Loudon, a tall, stringy officer whose insatiable thirst gave rise to the belief that he had been born with hollow legs.

At about that time a new station commander was posted in. His name was George Harvey – a charming man, but mild of manner and sober of habit when measured against most of his immediate subordinates at Coltishall. He had, I believe, been serving for some time in Air Ministry and he was obviously a little shaken by the way things were done on a fighter station. One evening, soon after his arrival, I had taken some Typhoons over to Coltishall for night flying practice. We were to have supper in the officers' mess before flying started and when I went into the ante-room I found the new station commander talking to Max Aitken by the fireplace. Max asked me what I would drink. It occurred to me that Harvey was not the man to approve of strong liquor before flying – particularly before night flying – so I said I would have a half-a-pint of bitter.

Max looked at me as though I was mad.

'Half-a-pint of bitter? Why on earth do you want half-a-pint of bitter?'

'Well you see, I'm flying later.'

'In that case you'd certainly better have a large glass of whisky. Never fly before drinking. Steward, a large whisky for Squadron Leader Dundas.'

I glanced at George Harvey. I had been quite right. He did not approve of strong liquor before flying.

Over the war years I saw quite a lot of Max Aitken, on and off. He could be a wonderful companion – one of those rare people of whom it could be said that life was never dull when he was around. I liked him and I admired him; and no doubt I was somewhat dazzled by his reputation as one of the rich young hell-raisers who had contributed much to the reputation of London's 601 Squadron before the war. His behaviour was often outrageous – but it was always amusing. He was a fine pilot and had great courage to complement his ability. In spite of his light-hearted manner, he ran a squadron as hard as any man I knew and was intolerant of weakness either in himself or others. He had commanded 601 Squadron at Tangmere during the toughest period of the Battle of Britain and had won a DFC in recognition of his own and the squadron's outstanding performance in the Battle. In due course he was also to be awarded a DSO. Although older than most of us, he went on flying operationally until the war was won.

In late October I went up to London with Max and Johnnie Loudon to attend a dinner party given at the Savoy hotel by an American flying association called 'The Quiet Birdmen'. They were on a visit from the USA and they had invited Battle of Britain pilots from every fighter station to the party. The sudden fusion of so many old friends in the West End of London produced a volatile atmosphere.

The guest of honour was our old chief, Sir Hugh Dowding. In his speech after dinner that night, delivered to a large assembly of exuberant and well-wined fighter pilots, he revealed for the first time in public his spiritualist convictions. All of us in that room felt the greatest respect and affection for our old chief, but his totally unexpected revelations, when he told us over the port that he was in regular communication with many of our friends who had been killed in action and that they were all in good shape and quite happy, had a macabre effect on the company. I am afraid that the reaction of most of us at that time was that 'the old boy had gone round the bend'. We were wrong, of course. Eighteen years later I spent some hours with Dowding when we participated together in a television programme to mark the twentieth anniversary of the Battle of Britain and it was quite evident that his exceptional powers of thought remained unimpaired. At the age of seventy-eight he appeared physically and mentally as spry as he had been at the time of the battle; yet his belief in communication with the dead remained unshaken.

However, the thought of our former colleagues lurking mysteriously around us in that room at the Savoy, waiting for us to get in touch with them, tended to charge the atmosphere and quicken our thirsts. Thus it was a collection of fighter pilots in an advanced state of alcoholic hilarity which was discharged at a late hour upon the night clubs of London.

Next day Johnnie Loudon, Max Aitken and I met for lunch prior to flying home to Coltishall in Max's twin-engined Oxford communication aircraft, which waited at Hendon. We lunched well and we lunched long. It was already dusk when we took off, after a brief stop for further refreshment in the old Auxiliary Air Force Officers' Mess at Hendon. The weather deteriorated as we went on our way. By the time we were approaching Coltishall it was pitch dark, raining hard and cloud was at a very low level. With anyone but Max in the pilot's seat I should have been scared stiff. As it was, I was merely suffering from an advanced state of anxiety. Johnnie Loudon, on the other hand, was insulated from all care – he was asleep in a folding canvas chair, tilted back against the side of the aircraft.

Every form of emergency lighting was switched on at Coltishall to help Max find the field and in due course we saw the reflection of this Blackpool-like display shining on the clouds ahead. Loudon slept soundly through the subsequent landing and was still asleep when the aircraft came to rest on the tarmac alongside the squadron hangar. Almost immediately the door of the plane was opened from the outside. Unfortunately the opening was at the very spot where Loudon's chair was leaning and our gallant wing commander flying came to life with a loud and startled oath on his lips, at the same time half-falling out onto the tarmac.

This undignified disembarkation would not have mattered too much but for the fact that the man who had opened the door was Group Captain George Harvey, the station commander. There was an unpleasant tenseness in the air as we were driven by Harvey to the officer's mess. Neither Loudon nor I could, I fear, have been described as entirely sober. Max alone was, as usual, entirely in command of the situation. Leaving the matter in his good hands, I departed with all speed for Matlaske.

At about this time I was ordered to give immediate consideration to the best ways of using the squadron offensively on air-to-ground and low-level air-to-air operations.

The sector of enemy-occupied territory closest to us lay a little more than a hundred miles across the North Sea. The whole Dutch coast, from the string of islands north of the Zuider Zee to those south of the Hook of Holland, was within range. It was up to group intelligence to provide us with information about likely targets, flak concentrations and the like. Meanwhile I decided that a good start could be made in familiarizing the pilots with some of the problems involved by flying a series of low-level sweeps up and down the enemy coast. There was always the chance of finding hostile shipping, which could be strafed effectively with our four cannons. And there was a reasonable hope of running into their escort aircraft.

I took off on the first of these operations on the morning of 5 November. Lieutenant Eric Haabjoern, Flying Officer Wallie Coombes and Sergeant Cluderay came with me. We flew out at a hundred feet, in order to keep below the German radar screen. Cloud was low and there were squally patches where visibility was reduced to a few hundred yards, the heaving sea was slate-grey and menacing. There would have been no hope for a pilot whose engine cut – too low to bale out and take to the dinghy, while ditching would be sure to end in disaster.

I flew a course designed to make landfall at the island of Walcheren. It was my intention then to turn left-handed and fly up the coast to a point north of Haarlem before turning left again for home. As we approached the Dutch coast we ran into a squall. Just as we emerged from it a line of ships appeared dead ahead, less than a mile away. There was only one thing to be done: fly straight at them, keeping as low as possible. The ship immediately in my path appeared different in outline from the others. I soon discovered the reason. It was a flak ship.

I had hardly taken in the picture when the lights started winking on this ship from stem to stern and a vicious concentration of tracer hosed across the wave-tops towards me. I opened fire in return and charged forward at a speed of four-and-a-half miles a minute, cannons thudding. In the seconds of our closing the persistent tracer streamed out at me from a cloud of spray and a galaxy of explosions where my shells were striking on and around the ship. Then I was upon her and I kicked the rudder to take me skidding low under her stern. As I flashed by there was an explosion behind me and my plane swerved violently, almost uncontrollably, to the left. It took most of my strength to stay on a straight course. I throttled back hard, as the flat coast line, studded no doubt with gun emplacements, loomed ahead. With decreasing speed it became a little easier to hold the plane. I found that it was possible to turn gradually – very, very gradually – to starboard against the pressure which all the time was trying to flick me round the other way.

Fortunately the flak ship had been near the back of the convoy, which was steaming north. And so I did not have to pass over any more ships as I gingerly worked my way round through 180 degrees. Settling on course, I saw in my mirror what had happened. An explosive shell, probably from a 40mm gun, had detonated against the tail fin and rudder, tearing a huge hole in one side of both. The resulting pressure caused by the slipstream was constantly forcing the back end of the plane round to one side. I knew that if I eased off on the rudder for an instant the plane would flick over and I would have about three seconds to live before crashing into the sea.

After a little while the strain on my right leg began to feel almost unbearable. I hooked the boot of my other foot under the left side rudder bar and pulled with that leg. Later I succeeded for a time in getting both feet onto the other side, but the resulting position was so uncomfortable that I could not maintain it for long. I debated with myself the possible advantages of baling out. But the ceiling was too low to make the attempt below cloud and if I climbed through before jumping it would then be impossible for my companions to follow my subsequent descent and plot my position. In any case I knew that even if I succeeded in baling out safely, which was questionable in the circumstances, I could not survive long in that cold, rough sea.

So I kept plugging on – some of the time I thought I was going to make it and some of the time I thought I was not. At last the low East Anglian coast came in sight and by good fortune I was smack on course. I was making for Coltishall, which was a little closer than Matlaske and better suited for an emergency landing. I called Haabjoern and told him I would go straight in and attempt a belly landing on the grass. He climbed ahead and transmitted this information to flying control.

I was determined to do as little as possible to change the attitude of my aircraft before landing, for fear of disturbing the precarious equilibrium I had achieved. No flaps, no wheels, and as little change of speed as possible – that was my landing plan. I was already at a low altitude and I eased the throttle very gently back to sink off more height gradually as I approached the airfield boundary. I just kept sinking until the ground skimmed by a few feet below. Then I slammed the throttle shut, cut the switches and waited for the law of gravity to take its course.

In the course of the whole war I think that I never sat so long so close to the doorway of death.

For the next few days the weather clamped down so badly that flying was out of the question. This was perhaps a good thing, for we had an operation of another kind on our hands.

Johnnie Johnson was getting married and I was to be his best man. His beautiful fiancée, Paula, was a Norwich girl and the wedding was to take place in that city. Johnnie's squadron had, a short time before, been removed to the frozen wastes of Caithness, where it had the task of protecting the Fleet in Scapa Flow during the winter months. But Johnnie took the precaution of coming south to stay with me at Matlaske two or three days before the wedding. There thus began a round of parties of which the memory has always been a little blurred. Johnnie beating Stanley Holloway to it by many years, said, 'for Gawd's sake get me to the Church on time' and then proceeded to pull out the stopper. Some forty-eight hours and innumerable whoppers later I fulfilled my part of the contract, watched the happy couple rattle off in a south-westerly direction in a borrowed car and sat back to survey the damage.

Perhaps the most noticeable aftermath of Johnnie's prematrimonial frolics in our midst was the fact that there seemed to be very little motor transport left intact. I had had a big Ford staff car, which suffered slight damage on the first night of the visit. That left a small Hillman car and a Standard van, not counting the camp commandant's privately owned Bentley. The next day, while returning from lunch, I rather carelessly turned the Hillman onto its side while cornering too fast on the taxi-track round the airfield. Later the same day Johnnie borrowed the Bentley, but returned on foot a few minutes later saying that, having failed to observe a bend on the road, he had landed up thirty yards inside a turnip field.

On the evening after the wedding some of my officers told me that they had arranged a meeting with some WAAFs at a nearby pub. They wanted to use the Standard van to collect the girls and ultimately take them home again. It was all very irregular, but I agreed to let them have the Standard on the strict understanding that they took a driver with them.

Next morning Basil Hudson, the squadron adjutant, came to my room while I was shaving. He looked very grave. He regretted to tell me that the Standard van had been smashed. Worse than that – the accident had taken place in the small hours of the morning and the back of the van had at the time been chock full of WAAFs who should have been home in their billets and tucked up in their warm little beds hours before. They had in fact got home with the milk, bruised and for all he knew battered, accompanied by a number of our officers whose strict sobriety was in question. The Queen WAAF, Basil concluded glumly, was on the warpath. He had the impression that she urgently wanted to find out by whose authority her tender charges had been driven out to drink the night away with a batch of evidently inebriated and probably lascivious fighter pilots.

The situation looked black indeed. When Basil gently suggested that things seemed to have gone a bit too far I was unable to disagree with him. There seemed every good reason to suppose that a natural and inevitable consequence of the night's work, coming on top of the general carnage which had recently occurred in the motor transport section, would be that before very long Acting Squadron Leader Hugh Dundas would be squadron leader no more. I faced the immediate future with no confidence at all.

But the day which dawned so blackly was to take a most wonderful and unexpected turn for the better. In the course of the morning I received a telephone call from John Grandy, at Duxford. He told me that a Wing of Typhoon bombers was to be formed forthwith in his sector. The first squadron, under command of my old friend Denis Crowley-Milling, was already assembling. A Wing leader was required and he wanted me for the job; 12 Group had agreed, but before finalizing it, he just wanted to make sure I was willing to take the job on.

Was I willing? All the morning I had been wondering gloomily how long it would be before my squadron leader's stripes were detached from my sleeve. Now here was the dazzling vision instead of a wing commander's stripes being substituted for them. Yes, by golly, I said, I certainly was willing, not to mention keen, and thank you very, very much indeed. I gave some guarded intimation of the reasons prevailing which called for speed, though I toned it down a bit, fearing that even the broad-minded Grandy might have second thoughts if he knew the full horror of the story.

He promised to fix it that very day. And by evening I had the signal in my pocket making me acting wing commander. In the overwhelming relief of escaping retribution for recent misdeeds I scarcely had time to consider the extent of the honour which had come my way – wing commander, equivalent to lieutenant colonel, and only four months after my twenty-second birthday.

I knew that I had been outrageously and undeservedly lucky. But then and for the rest of my life I adopted the view that when good things come your way, you should not allow the knowledge that they are undeserved to detract from the enjoyment of them.

CHAPTER THIRTEEN

The Tunis Campaign

My posting to Duxford as wing leader was all the more remarkable because there was in fact no Wing to lead. And it did not really look as though there would be one for some time to come.

Consequently I had a fair amount of time on my hands to think seriously about personal matters. Ever since 616 Squadron had moved from Kirton-in-Lindsay to Tangmere twenty-one months earlier my thoughts and activities had centred almost exclusively around the day-to-day interests, excitements and dangers of squadron life. I had developed a more or less single-minded absorption in that life and in the pursuit of war. When I had time off I either went home and savoured again for a few hours or days the atmosphere of my boyhood, or I went off in search of amusement with my friends in my own or some other fighter squadron. These forays were for the most part stag parties. They revolved around opening and closing time of bars. I had not often sought out feminine companionship and my relationships with women in general had been few and brief.

But there had been, and still was, one girl who was not a girl in a general way at all. She was for me a girl in a very particular way. It happened that when I went back to Duxford, Diana was stationed in nearby Cambridge, where she was a motor cycle despatch rider in the Mechanized Transport Corps.

For some time there had been a private understanding between us that we would one day be married. During those winter weeks at Duxford, when my duties were light and we were constantly together, the inclination to postpone matters no longer was very strong. I had a built-in resistance to the idea, because I knew that married fighter pilots seldom coped as well with the strain of their operational duties as did bachelors. I had seen a few exceptions, but I had also seen the point proved all too often. At that time my judgement of a man in my confined and circumscribed world was based simply on the answer to one short question: did he want to fight and go on fighting? I had known the best of them lose their will to go on after marriage. And I recognized well enough my own recurring weakness of will and spirit to know in my heart of hearts that getting married might lead me to seek the path to safety – a course which I strongly criticized in others. In retrospect it all seems rather smug and holier-than-thou; but that is the way it was.

But the winter, with its inevitable lull in operations stretched far ahead. At last the decision was made. Diana had no qualms – she was for taking our happiness while we could. My longing to do the same was overwhelming. I would fight my private battles against fear when the moment came. In the new year we announced our engagement and planned a wedding in February.

However, there was a man sitting behind a desk at Fighter Command who knew nothing about wedding plans, and no doubt cared less. He was the man who, in obedience to the demands of some higher and still more anonymous authority, had to provide the Command's share of the officers required for posting overseas. Some time around the middle of January 1943 this fellow found on his desk one morning an instruction to send two wing commanders for Wing leader duties with Spitfire units in North Africa. Running his eye down the list of possible starters whose name should he light upon but mine? There, he no doubt said to himself, is a chap without too much to do at present; there is a chap with extensive operational experience on Spitfires; and there also is a chap who has never yet stepped out of England in the service of king and country. Natually enough he wasted no time looking up the recent lists of wedding announcements published in *The Times*. He just reached for his pen and made a signal.

And in that way our marriage plans went up in smoke.

The second officer selected for this North Africa assignment was my old friend Johnnie Loudon. By good luck I got away before him and so was the first to present myself at air headquarters in Algiers, where Air Marshal Sir William Welsh, an officer who had served in Whitehall and in non-operational commands for the previous nine years, presided over the activities of the RAF in North West Africa. I was rather cast down when I reported to the senior personnel staff officer to find that no one really knew what to do with me.

'But I was sent out here at short notice to lead a Spitfire Wing,' I protested.

'Yes, I know that,' the staff officer replied. 'You see Gilroy, of 324 Wing, was slightly wounded. But it turns out that he's going to be fit again quite soon and we've promised him he can go back to his Wing. Don't quite know what's the best thing to do with you.'

I saw visions of being kept sitting in Algiers, three hundred and fifty miles behind the front line. That was not what I had been sent out to do. That was not why I had packed my bags in a hurry and postponed my wedding indefinitely. I sought out all the officers I knew in the headquarters and lobbied furiously for a posting to the front. After a short delay I was sent to 324 Wing at Souk el Khemis, as supernumerary wing commander flying. Poor Johnnie Loudon was not so fortunate. There was a limit to the number of supernumerary wing commanders who could be sent forward and he was kept kicking his heels in Algiers.

Number 324 Wing, where I reported on 3 February, had four Spitfire squadrons based on strips bull-dozed from the mud of the Souk el Khemis plain. The condition of these strips amazed and alarmed me – as it did all pilots arriving from England and accustomed to the idea that modern fighters could only be operated from well-equipped modern airfields.

Northern Tunisia is a land of rugged mountains separated by fertile plains. There were no permanent airfields between Bone, which was out of range of the front line, and Bizerta, which was in enemy hands. It had therefore been necessary for the tactical squadrons supporting General Anderson's First Army to make their own landing grounds. In the terribly muddy conditions which prevailed during the rainy season that winter it had been a job of extraodinary difficulty, the more so since the engineer units involved had no previous experience of that type of improvization.

It was a lucky thing that the officer commanding the forward operational units was Group Captain Ronnie Lees, a tough and determined Australian who had more than his share of the rugged individualism which characterizes his countrymen. I had known Lees for a long time. He had been one of Fighter Command's more outstanding pilots and commanders since the beginning of the war. And although I had always been rather afraid of him and had felt sometimes that his brusque, no nonsense manner indicated that he had no great liking for me, I admired him. This admiration was very much enhanced when I saw the difficulties he had overcome to keep his squadrons flying in Tunisia that winter.

The Spitfires were operating off strips of wire matting, laid on top of rushes which in turn had been laid on the mud. The strips were between eight hundred and a thousand yards long and only twenty-five yards wide. They were connected with the squadron dispersal areas by more strips of matting, laid down in narrow lanes. A pilot who put a wheel off the runway while landing – and it was all too easy to do so when coming down in a gusty cross-wind – was certain to capsize his plane. If you ran off the narrow taxi track your plane had to be man-handled back again. For the worst winter experienced in Algeria and Tunisia for many years had turned every square yard of ground into a quagmire.

Alongside these makeshift airfields the squadron officers and ground crews lived and ate in tents. The conditions in these cold, wind-swept, water-logged camp sites certainly provided a brutal contrast with the comfortable and solid quarters, mostly built in the middle and late thirties, in which all ranks of the RAF had been accustomed to live at home. Yet morale was high and it was obvious that everyone really took a secret pride in making the best of the foul conditions. I got the impression that Lees was not particularly pleased to see me when I reported to him in his office-caravan. He made it perfectly clear that Gilroy would continue as leader of 324 Wing. As I was there, I had better get an aircraft, get to know the squadrons and do what I could until Gilroy was fit. He took me over to the Wing mess, which was in one of the outbuildings of a big farm, introduced me to those of his officers who were there and left me to get on with it.

The squadrons in the Wing were Numbers 72, 93, 111 and 243. All but 243 had been famous numbers in Fighter Command. When I went round to visit them, one by one, calling in after dinner for a drink in the mess tent or driving my jeep up to their dispersal points, I was well enough received. But I quickly realized that they were not too impressed by anyone who had just come out from England. Of course, they had only been in Africa themselves since November, when Algeria had been invaded; but since then they had been bombed, strafed, submerged in mud and all the time engaged in tough air fighting. I was just another new boy from England. Furthermore I was not really their wing leader. And so they were polite, but a little distant.

Without delay I got myself an aircraft, which I gave to 243 Squadron to look after – partly because I thought the pilots of that squadron were a little less know-it-all in their approach than the others, partly because Tony Wehnman, an old comrade from Tangmere days – he had been in 145 Squadron – was one of the two flight commanders. I thought it wise to get flying quickly, before Gilroy returned, in order to establish my authority and position.

The first sortie I led was a particularly testing one. The Germans at that time were putting great pressure on the First Army front and it was vital for our troops to hold the line and indeed to make ground. If Anderson's army could break through to Tunis, the whole Afrika Korps, hard-pressed and pursued by Montgomery's Eighth Army to the south, would be denied an escape route and would be lost. And so a major German assault had been launched in the north-west against the point which seemed likely to be the most vulnerable in the allied line – the American sector, west of Kairouan. The first objective was the railway town of Tebessa, from which the German army could strike out north and north-west to encircle the British army and cut its communications.

The American troops in that sector were entirely without battle experience. It was tough on them that their initiation should have come at the hands of a numerically superior force of determined and desperate German veterans. In the circumstances it was neither surprising nor shaming that they should have given ground. The situation became very serious indeed and might have been fatal but for the fact that Anderson succeeded, in a movement executed with extraordinary speed, in getting an armoured division out of the line in the north, down through the mountain passes and into the battle in time to help the Americans stem the advance. It was to provide cover for their entry into the battle that I led 324 Wing on my first operational flight in North Africa.

As I led my two squadrons south through the mountains I soon began to wish that I had not entered into things quite so precipitately. The weather was bad, with a lot of broken cloud at different levels and scattered rain storms. I was totally unfamiliar with the terrain. Navigation was difficult, landmarks hard to pick out. A mistake could well have been fatal, for many of the mountain tops were hidden in cloud or obscured by curtains of rain.

Yet I could not climb to a safe height, for I had to keep the ground in sight if our purpose was to be achieved.

I was tempted to turn back. Good sense dictated that I should do so. Very likely the pilots flying behind me would have been secretly delighted, yet I knew that it would be damaging in the extreme to my authority and standing. I kept going and carried out the patrol as ordered, though without seeing any enemy planes. But then I had to get the Wing home again. The weather had steadily deteriorated and was really bad by the time I left our patrol line. The job would have been hard enough for a single pilot, able to twist and turn as he map-read his way up the valleys. I had twenty-three aircraft behind me and so had at all costs to avoid any violent manoeuvres.

Because of the mountains, radio contact with base was faint and uncertain. I could not rely on getting a homing bearing. In any case it was psychologically unwise to ask for one unless it became absolutely necessary; pilots liked to think that their formation leaders could get them home without such aids. And so we threaded our way through the murk and I prayed to the good Lord not to let me make any mistake. It was with a tremendous sense of relief that I finally saw the flat plain of Souk el Khemis slipping past beneath me, landing strips dead ahead. When I had taxied in and switched off my engine I felt as exhausted as if I had just returned from the fiercest of fights.

'Sheep' Gilroy came back from sick leave soon afterwards. I had known him slightly for some time, as he had been in the Auxiliary Air Force before the war, with the City of Edinburgh Squadron. His nickname stemmed from the fact that he was by profession a sheep farmer. He was a serious-minded man, sober in his habits, uncompromising in his judgement of men and their actions, outwardly dour, though he had in fact a quick and puckish sense of humour. If ever anyone nurtured a sincere desire to engage the enemy, regardless of his own safety, 'Sheep' Gilroy was that man.

Quite soon after his return the structure and establishment of the Command in which we were serving was changed. Ronnie Lees had been in charge not only of 324 Wing but also of 322 Wing stationed further back, at Bone, and of sundry army co-operation squadrons. The Command was now upgraded to an air commodore's post and Lees handed over to a senior officer. At the same time Gilroy was made commanding officer as well as wing leader of 324 Wing. This change in the order of things made my position easier, as Gilroy, faced with a certain amount of administrative work, now had rather less time for leadership of the Wing in the air.

The man posted in to command the group was Air Commodore 'Bing' Cross. He came up from Desert Air Force and so had very much more experience of mobile, close-support air warfare than the rest of us. He was only thirty-one years old with a fine operational record, which had begun with the leadership of a Gladiator squadron in the Norwegian campaign – a desperate and ill-fated venture from which he had been lucky to escape alive. He was bursting with energy and the will to get things done. His job was to prepare us for an all-out effort in early spring, when the ground would dry out and the Army's assault on Tunis would come to a climax.

I immediately developed a warm liking and admiration for Cross and I was delighted when he gave me duties of a definite and responsible nature. He made me a member of his personal staff with instructions that I should spend part of each day in the new group operations room, in the role of ground controller, and the rest of the time flying with the squadrons and leading the wings. Every evening I was to report to his headquarters to give him my views about the squadrons, the junior commanders, the pilots and the way the operations were being conducted.

This double life kept me both busy and happy. Throughout March and April the tempo of operations increased steadily, as the weather improved and the build-up for the final assault gained momentum. The Luftwaffe strength in Tunisia was limited and action was sporadic. But when it came it was hot and strong, for the enemy had the latest type of Messerschmitt 109Gs, as well as Focke-Wulf 190s. We were still equipped with Spitfire Vbs, which were markedly inferior in performance.

Early in April I was despatched by Cross on a very curious and unusual kind of mission. He told me that there was stationed at Bou Saada oasis, some two hundred and fifty miles south of Algiers, a French Air Force unit which had kept itself in splendid isolation ever since the allied invasion the previous November. Before that it had been technically under command of the Vichy-controlled Algerian government. My job was to fly down there, contact the commanding officer and, if I considered it desirable, invite him to get in touch with the RAF authorities with a view to getting Spitfires and coming to fight with us.

Because the flight involved a fairly long crossing of mountains and desert I decided to take another officer with me, in case of engine failure or accident. I also judged it wise to have a witness to my encounter with the mysterious Frenchmen. I selected Jimmie Grey, the Canadian commanding officer of 243 Squadron, and we set course for Bou Saada after lunch one day, in two Spitfires.

There was no sign of activity and little sign of life on the landing strip a mile or two north of the rich green oasis which, with its white houses and massive Beau Geste fort, lay like a bright splash of paint on the empty bare brown canvas of desert. A couple of small, old-fashioned planes stood by a tent in one corner. And as we circled I saw a man walk out of the tent and stand with his hands to his eyes, staring up at us.

I told Jimmie to go on circling while I landed and taxied in. I would call him if I wanted him to follow. With great caution – and a little trepidation – I landed and taxied over to the tent. The man I had seen ran towards me, waving and smiling. I called Jimmie and told him to come down. Our one-man reception committee was wearing the uniform of a lieutenant in the French Air Force. He was evidently astonished to see us, but he was courteous and friendly. I introduced myself and explained in my best French that I was a lieutenant colonel in the Royal Air Force; with my friend 'Commandant' Grey, I had come with a message from my general, which I must communicate personally to his own superior officer.

The Frenchman's astonishment became more evident as I spoke. However it soon became clear that he was more amazed that two such obviously juvenile officers should descend from the sky calling themselves colonel and major than doubtful about our credentials. He was unable to withhold comment on this curious phenomenon.

'Me, I've been a pilot for twenty years,' he explained. 'And look at me – still a lieutenant!'

He raised his eyes to heaven in silent protest against such clear injustice. Then he politely offered to drive us to headquarters to see his commanding officer.

From that moment events took on a flavour of fantasy. The commanding officer, a grizzled major well on in middle age, was courtesy itself. I did not want at that early stage, before seeing anything of the set-up, to say too much. My host, who, after all, had been sitting on the fence for several months presumably waiting to see who was going to come out on top in the African campaign, did not press me for the precise reason for my visit. He just went on sitting on the fence. But he pressed us both to stay the night as his guests. And he suggested that after going to our rooms for a while we might like to join him and some of his officers for a ride around the oasis on horseback – a suggestion to which I agreed, because I thought it polite to do so, though I felt some mild trepidation, as it was quite a long time since I had last been on a horse.

Dinner was a convivial meal, attended by about a dozen of our hosts. However much they may have wondered why we were there, they restrained their curiosity and behaved in the friendliest possible way. They expressed great interest in the performance of our Spitfires and some of them, as the wine went round and around, said how much they wished they could get some action. Maybe it was the wine working on me, but I decided that they were the sort of people we wanted with us and I told the commandant that I was authorized to offer them the opportunity to come and fight alongside us in the final liberation of Tunisia from the 'sale Bosche'. This information aroused great enthusiasm – maybe the wine was working on them too – and after more drinks it was decided that we should be taken out on the town, such as it was.

We were driven at lightning speed through the crowded and narrow streets and escorted into a white-painted single-storey house, where we were greeted with much courtesy by a middle-aged Arab woman. She escorted us into a large room and invited us to sit on cushions round the walls. I wondered what on earth was going to happen next. An old Arab man then walked into the room, carrying some kind of a flute-like instrument. He bowed politely all round and began to play.

An Arab girl came in and danced, followed by another, and another. They danced in a curiously stylized way, but with great grace. The Frenchman sitting next to me told me that Bou Saada was famous in the Arab world for its school of dancing. I nodded politely, but thought to myself that they would never make Sadler's Wells. Next came a young girl of startling beauty, who danced alone. When she had finished, the older woman who had shown us in whispered something to the commandant. He nodded. The woman next spoke to the old man with the flute, who went and stood in a corner, face to the wall. In this position he resumed his playing.

The girl who had danced last now reappeared. Only this time instead of her long flowing robe she wore nothing, absolutely nothing. She was a voluptuous creature and she danced in a voluptuous way, increasingly so as the tempo of the music quickened. The atmosphere in that little room became charged, as that desirable naked body gyrated before us in practised erotic rhythm. The music stopped and the girl slipped out.

Again the commandant spoke with our hostess. Then he crossed over to me.

'Madame's compliments, mon colonel,' he said, 'and the girl is yours for the night. You will be happy. She has great beauty, no?'

The vision of that voluptuous body swam before my eyes. She had great beauty, yes. And I would be happy, no doubt; but I remembered that I had come to collect a new squadron into our ranks, not to spend the night in the arms of an Arab girl from whom I would probably collect something quite different. To the undisguised amazement of my hosts I declined the offer.

No doubt someone else enjoyed the advantages of my refusal.

The Frenchmen joined us in time to take part in the last stages of the battle of Tunis. I flew with them on one of their early sorties but quickly decided not to repeat the experiment. They traversed the hostile sky with the joyous and undisciplined élan of a pack of sealyhams hunting a wood for rabbits.

The month of April 1943 was a glorious and memorable one for British arms in North Africa. The RAF played a vital part in the battle which led to the fall of Tunis and the surrender of German and Italian troops. By the time the last battle began the activities of the allied air forces had been co-ordinated by the creation of a unified command under Air Marshal Sir Arthur ('Mary') Coningham. His long experience in the use of air power in support of ground operations quickly made its mark. Even at my lowly level the change was soon apparent. Bing Cross was in almost daily contact and consultation with Harry Broadhurst, now an Air Vice Marshal commanding Desert Air Force in support of the Eighth Army in its advance from the south. New airstrips had been bull-dozed in the Souk el Khemis plain, now drying out in the warm sunshine, and most of the Spitfire squadrons in our theatre of the campaign were concentrated upon them. Light-bomber and night-fighter units settled alongside. A sense of ordered purpose replaced the atmosphere of muddled make-do-and-mend which had characterized our surroundings and our activities a few weeks earlier.

Much of the time we were engaged on close escort or area cover for the formations of medium bombers which now began to pound the German ground forces and their communications all day and every day. But the enemy fighter force, contracted into a small number of bases which themselves came under constant attack, did not offer any considerable opposition. We encountered them only occasionally. The German fighters in the area had at that time a job which was more vital even than the protection of their hard-pressed colleagues on the ground. It was the protection of the huge aerial convoys of Junkers 52 and Messerschmitt 323 troop planes which were engaged in a desperate endeavour to supply and reinforce von Arnim's armies.

The story of this extraordinary airborne supply line and its eventual destruction is, from the German point of view, both heroic and grizzly. During the months of February and March, when von Arnim's veteran divisions faced the total disruption of their normal lines of supply, the daily traffic of twin-engined Junkers and huge six-engined Messerschmitts across the sea from Sicily increased to an average of a hundred and fifty planes a day. The Junkers could each carry about two-and-a-half tons, the Messerschmitts ten tons. They were providing the German divisions with all their requirements for one day in every three. These lumbering convoys – the only thing which staved off the thrombosis of German forces – were watched and studied by the allied air commanders. And no one looked with more predatory eyes than Harry Broadhurst, but for a few frustrating weeks he was powerless to act as his fighters did not have the range to intervene.

By early April that situation had changed. Broadhurst's seasoned, battle-experienced squadrons – British, South African, Australian, Canadian, American and Polish – came at last within range of the sea routes traversed by the German air convoys. At first the efforts to intercept were unsuccessful. Broadhurst and his wing leaders persevered and their determination soon bore fruit. On 18 and 19 April units of the Desert Air Force sighted and fell upon large formations of the ponderous German planes as they approached Cap Bon in a mass. The carnage was so ghastly that our pilots had difficulty in giving coherent reports or estimates of planes destroyed. The unfortunate Germans were helpless, like sheep caught in a pack, swiftly broken up and massacred by hunting wolves. The only return fire came from the troops inside the transports, shooting with their sub-machine guns and rifles from the doors and portholes. The convoys were flying at very low altitude and the planes could not weave or manoeuvre violently without crashing into each other or into the sea. Some men dived from the doors, without parachutes, without hope, to escape from the inferno of flame and fire. Some German pilots just pushed the nose down and crashed into the water rather than wait for the final certainty of destruction by our fighters.

On each of these days more than half of all the German planes flying in the convoys were destroyed before our pilots ran short of ammunition. It was a shattering blow, but the enemy did not give up. Patrols of German fighters were put up off Cap Bon when the next convoys were due. Broadhurst quickly countered this move, he sent Spitfire squadrons from 244 Wing, his crack-fighter force, to engage the Messerschmitts and Focke Wulfs, while the slower Kittyhawks of his fighter-bomber Wings lashed into the transports. A formation of the huge Messerschmitt 323s was intercepted in the Gulf of Tunis by the South African 7 Wing, while 244 looked after the 109s above. The transport planes were loaded with petrol for von Arnim's trucks and tanks. Thirty-one of them fell, trailing fire across the sky, and burned on the calm surface of the sea.

It was the end, though still the desperate Germans did not altogether concede. If supplies dried up completely, then the whole vast fighting army in North Africa must collapse, without the slightest hope of being able to escape to fight again. It was a prospect which could not be accepted until every last endeavour had been made to avert it and the remaining transport planes were sent off at night. But there was an answer for that one too. The Beaufighters of 600 Squadron, commanded by my old friend Paddy Green with Peter Dunning-White as he second-in-command, flew out against them. So also did the Hurricane night fighters of 73 Squadron.

Stealthily, by the light of the moon, they finished the job.

From the other front we watched these successful activities of Desert Air Force with some envy. Those hunting grounds were beyond our range. We carried out some patrols north of Bizerta, in the hope of catching convoys coming in that way, but without result.

Allied air superiority over Tunisia was now absolute. There were enough squadrons and enough different types of planes to do every job – to intercept the airborne convoys and to bomb the supply ships which ventured within range; to strafe the roads and railway lines leading to the front and to batter the ports; to escort the medium bombers which prowled behind the enemy lines and also to maintain a constant curtain of fighter protection over the battlefield at Medjez. It was there that Anderson's First Army, now reinforced by armoured units from the Eighth Army, was operating under the strategic direction of General Alexander and had to break its way through the gap only three thousand yards wide before bursting out onto the plain, where all roads led to Tunis.

We played our part in all these operations, with diminishing opposition from the enemy in the air. In the end the greatest danger lay not in that old familiar menace, the Hun in the sun, but in the ack-ack shells of all shapes and sizes which were pumped up at us as we roamed behind the lines looking for targets of opportunity. The Luftwaffe could no longer sustain the fight. With only two bases left on African soil, and those constantly under attack, with the airfield on the island of Pantellaria bombed to destruction by the Americans, there was only one course left: to retire to Sicily and Italy and there await the next stage of the game.

And so the end came for von Armim's still vast forces on the ground. They could continue the fight no longer.

During three-and-a-half years of war my contact with the enemy had been always of the elusive, impersonal kind inherent in air fighting. The clash of glinting warplanes in the thin high air; the thud of bombs from an unseen plane; the taut swoop of firing fighter on anonymous groundlings and the vicious sparkle of answering fire; the patient stalking through the clouds of convoy-hunting bomber. These were the activities and the experiences which represented the meaning of war for me, personally.

That perhaps explains why it was a curious and almost solemn experience to drive out and see this hitherto impersonal enemy herded together in hordes of living, beaten human beings. I sat in my jeep and stared at them in their ugly uniforms and forage caps as I might have stared at animals in the zoo or men from another planet. There they were, in their tens and thousands – Hitler's invincibles, disarmed, dejected and looking pretty ordinary and harmless in defeat. There they were, the long and the short and the tall, as per the barrack-room song.

Well, fuck 'em all, I thought, taking the lyric one line further. They were the defeated ones, but they were safe. That was more than could be said for me and my friends. We had won; so we had to go on and start all over again.

Meanwhile, we settled by the blue waters of the Bay of Tunis to await developments. We swam and had our first taste of fresh fish for many months and visited the bars and other places of amusement or dissipation which Tunis had to offer. And we threw a series of parties for each other.

The Desert Air Force withdrew, together with the Eighth Army, to bases around Tripoli. But soon the word got around that we were to be incorporated into that legendary and incomparable force for the next surge forward, the attack on Fortress Europe.

While we were at Tunis a significant change was made in the command structure of the fighter Wings. The officers commanding were to be group captains, with wing commanders as second-in-command and wing leaders. 'Sheep' Gilroy was accordingly promoted to the rank of acting group captain and I awaited with confidence my posting to 324 Wing, with which I had generally flown during the African campaign, as wing commander flying.

When I received an invitation to lunch with Air Marshal Coningham at the magnificent villa he had occupied on the south shores of Cap Bon, I did not doubt that it was for the purpose of being briefed for the job. We lunched in the most glamorous and glorious surroundings I had ever seen in my whole life, except perhaps on a cinema screen. Among the other officers present were Air Vice Marshals George Beamish and Tommy Elmhirst, respectively Senior Air Staff Officer and Air Officer Organization. There also was Air Commodore O'Malley, the doctor who had finally passed me fit for flying training early in 1939. They were all extremely kind and friendly to me, but no mention was made during lunch of the purpose of my visit.

After lunch Coningham said that he would have his coffee outside and he asked me to join him. We sat on a terrace looking out to the south across the quiet blue sea. Here, I had been told, the Duke of Windsor had sat when he had rented the villa some time after his marriage. I thought I had never seen a more beautiful place and leant back at ease to enjoy the good coffee and the brandy which came with it – and, as I hoped, the news of my posting. When we were settled Coningham came to the point:

'You will have heard, I expect, that 324 Wing is joining Desert Air Force. It will be going to Malta in preparation for the invasion of southern Europe.'

I said this was what we were all expecting.

'And no doubt you would like to go with them.'

Hoorah! This is what I had been waiting for.

'Well, I'm terribly sorry to say that you won't be able to. We have to find another wing commander flying for that Wing.'

The sun faded and the glory of the afternoon departed. A cold chill ran through me as I carefully put my coffee cup down on the table beside me and asked:

'Why, Sir? I have been leading the Wing a lot, you know. I think everyone is quite satisfied with the way I have done it.'

'Everyone is entirely satisfied, including myself. But it's not as easy as that. I'm afraid that Broadhurst doesn't want you as a Wing leader. You know Broadie. He likes to choose his key men. And I have always let him do so.'

I remember that moment with photographic clarity. But why should Coningham's statement have been so desperately depressing to me? 324 Wing was off to the war again, after all, with all the dangers involved. Had not I for a moment, just the other day, felt a sense of envy for the German prisoners, for whom all risk of death in battle had passed? Well, if Broadhurst did not want me to lead a Wing, I could get a staff job and perhaps live in pleasant surroundings like these, with absolutely no danger at all.

The thoughts raced through my head, but gave me no comfort. They were pushed aside by the bitter realization that at this critical moment, when an operation of tremendous importance was about to begin, I was considered unsuitable as a leader in battle. I was much more than terribly disappointed. I was terribly ashamed. The arrogant contempt which I had developed for officers who would not play their part in battle, or were judged unfit to do so, was now turned inward upon myself.

I said, 'Do you mind telling me who the AOC has chosen for the job?'

'There isn't anyone else at the moment. Broadhurst is looking around.'

That made it worse. If there had been an officer available who had served closely with Broadie, someone he knew really well, then at least there would be some excuse I could give myself. After all, we had never even served on the same station, still less flown together.

I asked Coningham if he could give me any reason at all for the decision.

'I believe you once fell out with Broadie at some conference at Fighter Command. You took opposite positions over some important question of tactics. He feels it is better that his Wings should be led by people who see eye-to-eye with him.'

So that was it. My mind flashed back twelve months to a room at Fighter

Command ... Sholto Douglas leaning back in his chair saying: 'Well, Broadie, you have no experience of these planes, after all. I think we'll see if there's anything in what young Dundas says.' I remembered puffing myself up with pride and self-importance after that little victory. Now I could regret it at my leisure.

Coningham, I think, recognized and understood my distress. Anyway, he was very kind, he said he would not change the decision but he would do everything possible to find me a job as a wing commander somewhere outside Desert Air Force. Meanwhile there was something he wanted me to do right away. Two Typhoons had arrived at Casablanca by sea which had to be flown to Egypt for flight trials in sandy conditions. The officer who had been sent out with them was ill. He had been accompanied by a Sergeant Pilot, named Myall, who had been with me in 56 Squadron and had got the two planes as far as Algiers. Coningham said that I was the only experienced Typhoon pilot in his command and asked me to go to Algiers at once to fly one of the planes to Cairo. Myall was to accompany me in the other one.

I went off to Algiers in a mood of black despair and there, in the company of two friends in the Army, I proceeded to paint the town very, very red indeed. Next day, feeling close to death and assuming that this was due to my excesses of the previous night, I clambered into the cockpit of one of the Typhoons armed with a small-scale map of the Mediterrranean which I had torn off the wall of a wash-room at the airfield, and set course eastward, Sergeant Myall following trustingly behind.

Our first refuelling stop was Bone, three hundred miles along the coast. About half-way there I began to wonder whether I was going to make it. I felt so ill that I seriously feared I would pass out. But there was nowhere to land on the way, nothing beneath us but jagged mountains running down to the coast. Somehow I held on as far as Bone, landed and taxied in. When I climbed out of the Typhoon and stood beside it the world went round in circles and I passed out. I was taken to someone's tent and lay down shivering on a camp bed with a high temperature. Next morning my temperature was down and though I felt weak at the knees I knew I was better. I never shall know whether that was the worst hangover I ever had, or something quite different. Anyway, having equipped myself with some more adequate maps, we went on our way.

I spent a comfortable night at Tripoli with John Grandy, who was commanding a group there and living in a magnificent though ostentatious palace which had been used by some Italian nabob during the years of the Fascist occupation. We reached Cairo safely the following evening and I set out both to enjoy the seven days leave which Coningham had told me to take and also to do a little lobbying among any influential friends I might find in Middle East Headquarters.

CHAPTER FOURTEEN

Back into Europe

One of the first people I met in the bar of Shepherd's Hotel was Air Commodore Richard Atcherley.

The Atcherley twins, Richard and David, were two of the most extraordinary characters in the RAF. They combined an unrivalled capacity for unorthodox behaviour with a dedicated sense of duty and love of the Service. It requires a whole book to tell the hilarious and hair-raising story of their careers, and such a book – *A Pride of Unicorns* – was written many years ago by the poet John Pudney.

At that time, Dick was commanding the organization in Desert Air Force which was responsible for planning and controlling the day-to-day operations of the Wings and squadrons. His brother David, who had been in charge of night fighters in Algeria and Tunisia, had been badly injured in the course of a typical Atcherley escapade. After the subjugation by bombing of the Italian-owned island of Pantellaria he had been unable to resist the temptation to go and have a look at the place, even though it had not yet been formally occupied by allied troops. Accordingly he set off in a Beaufighter which he attempted – very foolishly, no doubt – to land on the damaged runway. Inevitably he and the Beaufighter ended up down a hole.

Finding himself stranded, David explored a damaged hangar. There he found a small Italian aircraft, which looked serviceable and which furthermore, contained petrol. With great effort he pushed it out onto the runway and lined it up along a clear path for take off. Finding no self-starting arrangements, he decided to swing the propeller by hand. While he was making the preliminary adjustments the engine fired prematurely and the propeller smashed down on his shoulder, injuring him severely.

Over our enormous glasses of that unique purple-coloured libation know in Shepherd's as a 'suffering bastard', Dick told me this sad story. I then told him my sad story, pouring out to him the sense of shame and desolation which I felt. I do not know whether he had anything to do with it, but a couple of days later I received instructions to return forthwith to Tunisia and to report again to Coningham's headquarters. There I was told that I was to go to 324 Wing as acting wing commander flying, pending the appointment of a permanent replacement in due course. Coningham hinted that, possession being nine points of the law, he thought my chances of holding the job were good.

On these uncertain conditions I joined 324 Wing on 11 June. The advance ground party was already on its way to Malta and on 12 June 'Sheep' Gilroy flew off with the first squadrons, leaving me to follow with the remainder the next day. The four original squadrons of the Wing – numbers 72, 93, 111 and 243 – had now been joined by a fifth, Number 43. This unit was commanded by an old auxiliary friend of mine, Michael Rook, whose family had been co-founders of the then well-known Nottingham wine merchants Skinner and Rook, always referred to in the RAF as 'Skin 'em and Rook 'em'. I decided to keep my Spitfire with his squadron.

Malta at that time really lived up to its description as a floating aircraft carrier. Twenty Spitfire squadrons, five night fighter squadrons and an assortment of reconnaissance and transport aircraft were crowded into its three almost contiguous airfields. And with this armada of planes there descended upon the little island, and upon its bars, beaches and female population, a migratory flock of volatile airmen, determined to make the best of the island's comparatively civilized amenities before pressing on into the unknown.

We had fun at Malta, but we also flew hard. The British Eighth Army and American Seventh Army faced a momentous task. For the first time since the British had been sent packing from France and then from Greece in the spring of 1940 an allied army stood ready to invade Europe. The sea crossings were long. None of the troops, British or American, had experience of an amphibious invasion against a defended enemy coastline. And Sicily was an island of rugged terrain, favouring defence. In these circumstances a heavy responsibility fell upon the allied air forces. Mere air superiority was not enough. It was necessary not only to dominate the sky above the invasion beaches but also to disrupt the enemy's lines of communication throughout Sicily and far back into Italy.

The Spitfire squadrons based on Malta had a double role. They were called on to escort the twin-engined bombers which flew daily, several times a day, to attack the ports and communications systems in Sicily and at the same time to bring the defending fighters to action. And although the allies enjoyed a very considerable degree of overall air superiority, taking into account medium and heavy bombers, there was in the early stages of the campaign something like parity in the fighter field. And that, according to the book, was not enough for an invasion. But there was an absolute limit to the number of fighters which could be squeezed onto Malta. The Americans achieved a miracle by building a landing strip on the hilly and rugged island of Gozo, just north of Malta, in exactly ten days. But when that airfield was manned there were still only about twenty-five day fighter squadrons within operational range of Sicily – say four hundred and fifty planes of the front

North Africa and Italy, 1943–45

line. Against this force the enemy mounted about two hundred and sixty German and two hundred Italian fighters in Sicily and southern Italy.

We had some hard fighting during those last days of June and early days of July. At least once a day and sometimes twice, I led the Wing across the sixty miles of sea to patrol over Sicily. My squadron commanders were all experienced and aggressive men, notably 'Danny' Daniels, who led 72 Squadron, and a rough, tough Canadian called George Hill, who led the famous 'Treble One'. Both these characters were pugnacious in the extreme and Hill, particularly, was, I sensed, rather suspicious at first of the restraint which I insisted on imposing on individual enterprise. His purpose was always to get stuck into the enemy at the earliest possible opportunity, plane for plane and man for man. He was impatient of any restraint which might seem to him to run counter to that objective. But I knew that a Wing formation was only fully effective if held together under tight control of the leader so that he could direct his squadrons and flights into the battle at the critical moment, under favourable circumstances and in an orderly manner. In this way the thirty-six planes in a Wing could not only impose the maximum damage on the enemy, but also on occasions, be able to reform as a unit after a fight and so remain an effective force both for attack and in defence.

My suspicions about Hill's feelings were confirmed one evening at a party given by Broadhurst for all the officers of 324 and 244 Wings. The drink flowed freely and George Hill got well primed up for an argument. Late in the evening he told me rather truculently that he did not agree with my way of doing things. He said there had been times when he reckoned he could have chased off and got in a shot, but I had held him back. What kind of way to fight was that? He had reached the stage when he was not in the least worried about what he said and he was not to be shut up, though fortunately he managed to stay on the right side of the line which divides criticism from insult.

The next afternoon there was an operation in which our job was to sweep across the southern half of Sicily above and behind a force of bombers attacking a target in the south-eastern part of the island. We were heavily engaged by a large force of Me 109s and had a long hard fight, in the course of which the Wing was broken up into small formations of two and four Spitfires, fighting individually.

As I was working my way back towards the south coast, accompanied only by my number two, we were attacked and my companion's plane was damaged. Luckily we were by that time crossing out over the coast and the enemy planes pulled away. I saw a thin trail of white smoke streaking out from the Spitfire beside me and called my number two to tell him that his cooling system had been punctured. He replied that his engine temperature was already rising. We were still at about sixteen thousand feet, so I thought there was a good chance of getting him back to Malta. I throttled back, told him to follow close beside me and concentrated on steering the best possible course and on flying in such a way that we would cover the distance as quickly as possible but with the minimum use of power.

When we were half-way home George Hill came up on the radio to say that he was circling a dinghy with a man in it two or three miles south of Cape Passaro. He asked for company. I pressed the button on my petrol gauge – only about twenty-five gallons left. Then I asked George how much fuel he had. He replied that he had twenty-five gallons and added that some 109s had put in an appearance.

By this time Malta was in sight, some fifteen miles ahead. I called control, passed on George's message about the dinghy and asked them to get the Air Sea Rescue boys to work. Then I told my number two to carry on alone and ordered him to bale out if his temperature rose above 125 degrees. He would be picked up without difficulty.

Swinging round to the north again, I dived for Cape Passaro, asking George for his exact position and altitude. He replied that he was under five hundred feet and that six 109s were circling above, taking it in turns to make a pass at him. He added that they were a shower of bloody bastards.

'Hold on,' I said, 'I'll be with you in a couple of minutes.'

By good luck I saw them from a distance – a solitary Spitfire circling low over the water and a ring of Messerschmitts flying round a few hundred feet above. As I watched, one of the German planes detached itself and curved lazily down to the attack. The Spitfire turned sharply in to meet it and for a moment the two whirled round together, before the Messerschmitt zoomed up again to join the others.

I thought: 'Well, this is a pretty damn fool situation; we'll both be out of gas before we can extricate ourselves from this lot.'

Shouting to George that I was now among those present, I opened my throttle wide, leant the nose down and charged in and through the circle of Messerschmitts, taking a couple of snap shots as I went. Then I hauled up and round and came in for another whirl through. My hope was that by behaving in the most dramatic way possible I would confuse the enemy force into thinking there was more than one of me. It was a pretty forlorn hope, but it seemed to work. George meanwhile was climbing up into the *mêlée* and adopting a pugnacious attitude. To my relief and utter amazement the 109s withdrew to the north.

We were both dangerously short of petrol and we flew back towards Malta as delicately and gingerly as cats on hot cinders, conserving every pint. Ten miles out I asked control to clear the circuit. My gauge showed just under five gallons. George said he reckoned he could make it. We crept down onto the runway and by the time I reached dispersal point my tank was dry.

'Thanks, Sir,' said George.

'Don't mention it, old boy.'

The argument of the night before was forgotten, by tacit agreement. From that moment on, George Hill was a staunch friend and supporter.

The Air Sea Rescue boys picked up that pilot and took their launch in within range of the enemy coast to do so. And when they got him in the boat they found he was a German – and a rude and angry German at that. I am not surprised that he was angry. But if I had been there and he had been rude to me I would have chucked him back in – without his dinghy.

Activity, in the air and on the ground, became hectic during the days and nights leading up to 10 July. In the air the final stages of the softening up for invasion went on round the clock. On the ground we extracted as much enjoyment as we could from our last days in Malta. There were dances and parties every night.

I had spent a good deal of time while on the island with a nurse from one of the hospitals. She had told me that she was unofficially engaged to an army officer who was on a commando course near Tripoli. Well, Tripoli was a long way away over the sea, and what the eye does not see the heart does not grieve over, so I was able to put the brave fellow out of my mind. But one night he turned up. It was at a dance, where I had arranged to meet my nursing friend. Seeing her come into the room I went over to greet her. As I approached she grimaced at me, evidently trying to make me understand some warning. Standing immediately behind her, glaring around him like a ferocious bull, was the toughest looking character I have ever seen. The light of battle was in his eyes. He could have eaten two of me before breakfast. Discreetly, I faded and subsequently spent as much time as possible in the air safely out of reach of any commando tricks.

On 10 July, the morning of the invasion, Micky Rook and I, who shared a room in a little house by the sea south of Hal Far, were called long before dawn. Outside the wind was blowing fresh and gusty and there were white-capped waves on the usually quiet calm sea. As we shaved and dressed and sucked down our tea we wondered anxiously how the armada of little landing craft converging on hostile Sicily from south and west was faring on those rough waters.

The Wing had been allotted a particular stretch of beach to protect, south of Syracuse. It was our job to maintain a patrol over this area throughout the day, in squadron strength. 'Sheep' Gilroy had decided to lead the first patrol, taking off at first light. I was taking Micky and his squadron to relieve him. Each squadron was due to spend thirty-five minutes on patrol and I had given strict orders that formation leaders were to start for home as soon as their time was up, provided their relief had arrived and they were not engaged with the enemy. The observation of these orders was import-ant, if the day's programme was to be properly carried out.

There was an unforgettable atmosphere of tension and excitement at the dispersal points that historic morning. The Spitfires of five squadrons stood ready, silhouetted against the lightening sky, engines crackling to life and exhausts showing red as the mechanics warmed up the engines and tested the magnetos. I visited each squadron and had a word with the COs and their pilots before going back to see Gilroy and his squadron taxi out and take off.

Twenty-five minutes later, as the sun popped up from the sea in the east, it was time for me to get going.

'Well, Micky, all set?'

'All set, Sir.'

'Let's go, then.'

Micky, as always, walked out to his plane wearing a pair of floppy bedroom slippers. He was an enormous man and his feet were so big that he found it more convenient to leave his flying boots permanently wedged in the rudder bars of his aircraft. And so when this genial giant went flying off to war he was invariably to be seen handing a pair of slippers to his ground crew before climbing into his Spitfire in stockinged feet.

We crossed great convoys of landing craft and supply ships churning through the wind-flecked sea. Even from twelve thousand feet it was possible to see that the waves were breaking hard against and, often, over them. I felt sorry for the men who had spent the hours of darkness squatting miserably in those pitching, rolling little boats, wet and cold from the spray and the wind, with nothing to do but contemplate their arrival on a hostile beach. I thought that our way of fighting a war was not so bad. Probably there was danger just ahead, but we had risen from comfortable beds, with sheets and pillows, and we could hope and expect to go back to a hot breakfast far from the sight and sound of battle.

We reached our patrol line on time and I reported my arrival to Gilroy, asking him if he had seen any enemy planes. He told me that all was quiet. I could not believe that the quietness would last for long. Beneath us our ships were everywhere, landing craft were still chugging into the beaches, looking like swarms of little swimming beetles. Cargo ships stood off-shore, surrounded by smaller boats taking off supplies. Destroyers rushed back and forth like sheepdogs flanking their flock. We were watching the mightiest amphibious military operation yet undertaken in the whole history of arms. Surely the enemy air force must react strongly within the next few minutes. There were targets galore. I settled on patrol and awaited the onslaught.

Our patrol line was immediately above the strip of coast where British parachute and glider troops had landed during the night. There were sad signs that this operation had ended in disaster for many of the men taking part. The wreckage of many gliders could be seen at the water's edge. Some were practically submerged. I was looking at the evidence of a miscalculation which resulted in bitterness and recrimination.

The British gliders had been towed in by American Dakotas. Many of the gliders were cast off too far from the coast for them to reach their designated landing areas in the face of the strong west wind which was blowing. Quite a lot crashed into the sea. Later, the British survivors suggested that the fierce anti-aircraft fire which had come up from the Syracuse area had frightened off the Dakota crews. The Americans insisted that they had been guilty only of a miscalculation arising from the fact that the wind was much stronger than they had been briefed to expect. Whatever the true explanation of this unhappy incident may have been it is fair, in telling it, to add that if the American pilots did indeed flinch from the flak, then they were very untypical of their breed. I escorted many, many daylight raids by American bombers and again and again I came home filled with admiration for the way they flew unflinching through the barrage of fire, every plane holding perfect formation, swerving only to close the ranks when one or more fell away in smoke and flame.

Many minutes after I had reached our patrol line Gilroy was still there with his squadron. I called him again to make sure that he knew we were there and he replied that he intended to stick around a little longer. Typically, he was determined to be in on any fighting which might come our way. However, the expected reaction did not materialize before Danny Daniels called me to say that he was there with 111 Squadron. That made three of us. I acknowledged Daniels' message and said that I was returning to base. As I was landing Gilroy called to say that he was entering the circuit with his squadron, some of whom were short of petrol. I ordered the rest of 43 Squadron to hold off so that the others could land first. Standing by my plane, I watched them come straggling in. One was approaching from the sea to the south of the airfield. Its engine was dead and the pilot was evidently trying to stretch his glide. As I watched he made a slight turn to line up with the runway. But he was flying too slowly. The turn caused his plane to stall. It flicked over in a spin and fell into the sea.

All that day and the next we flew hard over the beachhead. Enemy reaction developed, with bombers, fighter-bombers and fighters striving to disrupt the allied advance while the German army was deployed at top speed to block General Montgomery's thrust towards Messina. The enemy air commander threw in formations of old fashioned Italian Machi fighters and fighter-bombers, as well as his more formidable Me 109s, which at that stage appeared both in German and Italian colours. But at all times the Spitfires were in numerical superiority and we kept well on top of the battle, scoring many victories.

On 11 July – D plus one – we were told that we were to start moving that evening to Comiso, an airfield with a long concrete runway about half-way along the south coast of Sicily. Our advance ground party had landed only a few hours after the first wave of invasion troops and was waiting to receive us. And so that evening, before taking off for the last patrol, I stuffed a pair of pyjamas and a sponge bag down beside me in my cockpit and ninety minutes later put my wheels down gingerly on the soil of Europe. It was a moment to remember and a memory to treasure as I taxied cautiously in, guided by an airman who had run out and jumped up to sit on my wing, I was delighted to see that the periphery of our new airfield was littered with enemy planes, many of them brand new-looking 109Gs. And that, I said to myself, feeling something of a conquering hero, makes a welcome change as a way of meeting a gaggle of bloody Messerschmitts.

Spirits were high in 324 Wing as we settled in at Comiso, and deservedly so. We had joined Desert Air Force as new boys, on probation. The older units of that incomparable force had an experience of mobile air warfare in support of an army which was quite unique, their achievements were glorious and they were justly proud of their traditions. They had regarded us with some suspicion, as interlopers, white at the knees and green in the matter of experience. But in the three weeks leading up to our landing in Sicily we had done better than any other Wing in the force. We claimed sixty-three enemy planes destroyed and many more probably destroyed or damaged. Those were figures which put the new boys way up at the top of the form and which made me very proud and happy. And they were figures, also, which Broadhurst did not ignore. My gratification was complete when he told me that he was no longer looking around for a new leader for 324 Wing. The shadow which had been hanging over me since my meeting with Coningham six weeks earlier was thus lifted and dispelled.

Even in captivity the Me 109 turned out to be a dangerous machine. And many of the abandoned enemy planes found at Comiso had been damaged to a greater or lesser extent. But two or three were in almost perfect condition. We decided to prepare the best of these for flight, thinking not only that it would be fun to fly one of the German planes which, over the years, had given some of us so many hard fights, but also that we might learn something useful from the experience.

After a couple of days the engineer officer in charge of the project told me that the Messerschmitt we had selected was ready to go. It had been painted bright yellow and was conspicuously adorned with British markings. I said I would fly it at six o'clock that evening and ordered the duty operations officer to advise other Wings that there would be an entirely friendly Me 109 in the Comiso circuit at that time.

The news of my intentions got around and when I went down at the appointed hour I found most of the Wing pilots and ground crew sitting around the perimeter track waiting to watch the fun. I strapped on my parachute and climbed into the unfamiliar cockpit, rather wishing that I had not committed myself so publicly to the enterprise. However, it was too late to draw back.

I found the cockpit unpleasantly cramped. The seat was positioned in such a way that the pilot was, so to speak, sitting on the floor in a semi-recumbent attitude, legs sticking straight out in front. Such a position had its advantages, because a pilot whose legs and head were as nearly as possible at the same level would not black out as quickly as one who was seated in an upright position. But the unfamiliarity increased my anxiety as I sat listening to the engineer officer explaining the functions of the various knobs and dials. Futhermore, it was very evident that the 109 had not been tailored for pilots of my height. When I closed the cockpit canopy I found that it pressed down on the top of my head; and when I turned my head to right or left my long nose was liable to come in contact with the perspex hood.

The moment arrived when I could hesitate no longer. I pressed the starter button and, to my secret disappointment, the engine fired straight away. Gingerly – for the forward view when taxi-ing was deplorable – I crept round the perimeter track, past the crowd of grinning specators, to the end of the runway. Then I turned across wind, ran up the engine to half-power and tested the switch which controlled the pitch of the propeller, changing it as desired from coarse to fine, or vice versa – the equivalent of high and low gear in a car. As I did so, I watched the engine revolution counter. The needle did not move. I flicked the switch again, listening intently to the note of the engine: still no change. Evidently the pitch control was unserviceable. In the circumstances it would have been crazy to take off. I taxied back, secretly relieved, but trying to look disappointed. The spectators made derisive signs as I passed.

A few days later the engineer officer again declared the plane servicable. As the Wing had by that time moved from Comiso and I did not want to go back there myself, I authorised one of the more experienced pilots of 72 Squadron to fly the Messerschmitt. He took off successfully, with an escort of Spitfires, and climbed to five thousand feet over the airfield. Then without any warning, the engine caught fire. Fortunately the pilot escaped by parachute and landed unhurt. I said a few silent Te Deum's and put an embargo on any further tinkering with enemy aircraft. The exception was a neat little Italian bi-plane, with two open cockpits, which had abeen appropriated by 72 Squadron. This uncomplicated machine was not likely to break any bones. It did good service as a run-about for a long time to come.

Comiso, with its long concrete runway, was wanted for the twin-engined medium bombers and we had moved after only three days to Pachino, on Cape Passaro, the most south-easterly point of Sicily. Our landing strip there had been bulldozed out of a vineyard and was only just over eight hundred yards long. The 100 Spitfires of the Wing were dispersed among the vines along narrow taxi tracks prepared by the earth-moving equipment of the Eighth Army's Airfield Construction Regiment. Whenever a plane moved, its slipstream raised a thick cloud of fine red dust, which became a permanent feature of our lives, permeating our bodies, sprinkling our food, penetrating our bedding.

The Wing headquarters tented camp site was on a small hill overlooking the landing strip. It was a bare spot, with only a few olive trees to provide some shade from the fierce sun of high summer. Within a few hours of our arrival it was infested by more flies than I ever had the bad luck to encounter anywhere before or since. Disgustingly they crawled and buzzed over and around us, in the mess tents, in the field kitchens – and in the latrines. They fastened greedily onto any cuts or scratches, so that within a few days half the members of the Wing had painful, septic patches on arms and legs. They were an even greater curse than the mosquitos, which abounded in Sicily and caused a greater and graver malaria epidemic than ever the Eighth Army or Desert Air Force had suffered in Africa.

In these less than salubrious surroundings I was sitting outside my tent one evening, a little before dusk, drinking a bottle of beer amd smoking one of the vile 'V' cigarettes which some never-to-be-forgiven evil genius of a staff officer had caused to be manufactured in India for distribution, through NAAFI, to British forces in the Mediterranean area. The quietness of the evening was broken by the sound of many engines and I saw a big four-engined American Liberator appear overhead at a low altitude. It circled the landing strip and to my amazement I saw the wheels come down as it turned and lined up on the 'runway'.

The idea of a Liberator, which had – for those days – an exceptionally high wing-loading and required a very long runway, trying to land on our little dust patch was in direct contravention of all known theory of flight. Of course the pilot would open up his engines and fly on to land elsewhere. I watched hypnotized as the great plane lost height and approached the end of the landing strip. A crowning touch of craziness was added to the scene by the fact that the approach was being made down wind. A barrage of red Verey lights shot up from the control officer's tent, half-way along the strip. But the Liberator pilot was not to be deflected. He skimmed over the end of the strip travelling at about 140 m.p.h., four or five hundred yards later his wheels touched. Several tons of metal, travelling at high velocity, shot along what remained of the strip and ploughed into the vineyard beyond. There was an explosion of dust and an expensive sound of rending metal. I closed my eyes and waited for silence.

The next few minutes produced utter pandemonium. Several more Liberators flew down from the evening sky and descended upon us. Each successive landing was more spectacular than the one before, as the pilot swerved violently after reaching the end of the runway in an effort to avoid his predecessor. They came from both directions, though luckily never simultaneously. About a quarter-of-an-hour after the first plane had appeared there was a crescent-shaped collection of Liberators at both ends of our runway, all in undignified positions, some resting on one wing tip, some flat on their bellies with both wheels collapsed, some still standing on their undercarriages deep in the vineyard, like elephants lost and far from home.

While all this was going on I had been dodging about in my jeep, keeping a wary eye open for descending planes. The Liberator crews were in a state bordering on shock, their answers to my questions were inarticulate. They had evidently been subjected to exceptional and excessive strain. Some of the planes were pock-marked with bullet holes, some of the crewmen were wounded. Those who were whole and unharmed hugged each other in the joy and relief of being safely down on the ground. I saw some patting the ground. Others put handfuls of red Pachino dust into their pockets.

Eventually I discovered that our visitors were survivors from the dramatic raid undertaken by a large force of American planes against the oil fields at

Ploesti, in central Rumania. It is not surprising that these men were in a shocked condition. They had suffered a terrible ordeal that day, since taking off from Egypt early in the morning. They had been engaged on a twentieth-century version of the charge of the light brigade, except that their valley of death had been several hundred miles long and had involved a two-way journey. For hours on end the gallant Americans had plugged on through the hostile sky, constantly attacked by fighters and hammered by flak. The raid, boldly conceived, had been badly organized. I was told that individual crews had been sent off on this long-range mission relying entirely on the navigators in the leading planes. If and when those leaders were shot down the men behind found themselves deep in hostile territory with inadequate knowledge about where to go or how to get there. The few survivors who struggled back from this terrible ordeal had only one idea in mind: to get their damaged aircraft down to earth on friendly territory before their fuel reserves, already critically low, ran out altogether.

For the Spitfire squadrons of Desert Air Force there was more discomfort than danger in the Sicily campaign. The Germans made no attempt to hold the whole island, but quickly concentrated their forces in front of Montgomery's Eighth Army in order to defend the east coast port of Catania and the approaches to the Straits of Messina. The terrain greatly favoured the defenders, with the ten thousand feet high mass of Mount Etna standing in the way of the allied armies, its lower slopes falling steeply to the sea on the east side and forming rugged valleys and gorges to the west and north-west – stiff natural obstacles to the advancing allied land forces.

The job of the tactical air forces was to blast away at the enemy's defensive positions in order to open up a way through for the troops on the ground. And so there was plenty of hard and dangerous work for the fighter-bomber Wings, while the Spitfires were engaged almost entirely on escort work. As the campaign proceeded there was less and less reaction from enemy fighters. In an effort to make up for this lack of air power the Germans greatly strengthened their anti aircraft defences. In their 88mm gun they had an all-round weapon which was used with great effect against aerial as well as ground targets. These guns were packed into the valleys at strategic points and it was around the base of Mount Etna that we made our first close acquaintance with them. It was to continue throughout the Italian campaign at the cost of the lives of scores of Desert Air Force pilots.

Whether by coincidence or design – the former I assume though you never could tell with Montgomery – it was on 3 September, exactly four years after the declaration of war, that the commander of the Eighth Army launched his assault across the Straits of Messina to gain a toe-hold on the European mainland. On that day I led the Wing on covering patrols, watched the little ships bustling busily across the narrow waters and saw the sand-coloured trucks and jeeps of the Eighth Army fanning out along the roads which led, ultimately, to Hitler's Germany.

And I looked back, too, across the four weary and hazardous years to that day at Doncaster when a carefree boy had secretly revelled in the prospect of the fighting to come and had fervently prayed that it would not all be so quickly over that he would not get into it before his training was complete. Now I was just twenty-three years old with the leadership of five squadrons of fighters and I felt I had done enough fighting to last me a lifetime. I was tired and feeling the strain – my only relaxation was in drink; if the war had stopped then and there, no one in the whole world would have been happier about it than I. I was very conscious of the fact that I was – or at least believed myself to be – inadequate and inexperienced in almost every respect save that of being a fighter pilot. I still felt myself a gauche schoolboy by comparison with most other officers of my rank who had enjoyed a little time in which to grow up before the war.

There was little time for introspection and still less for writing down my thoughts. We were moved to a new strip which had been prepared for us on the north coast of Sicily, about fifty miles west of Messina, and placed under operational command of American Air Headquarters, though for administrative purposes we remained part of the Desert Air Force.

The object of this move was soon made known to us by our new commander, Brigadier 'Shorty' Hawkins, of the US Army Air Corps. He paid us a visit to make our acquaintance and to inform us that we would be operating in support of a landing by the American Fifth Army, a mixed invasion force of US and British divisions. The landing was to be in the Bay of Salerno. I took a look at the map and was very unhappy to find that Salerno was one hundred and seventy-five miles from our base – and every mile over the sea. That meant that our planes would have to be equipped with long-range fuel tanks; and they would not be the normal thirty gallons tanks which we often carried on operations, but the big ninety gallon version, which made a Spitfire a cumbersome travesty of its normal self and which had a habit of not dropping off when you tried to jettison it – a very necessary preliminary to combat.

By and large I did not like the look of this Salerno caper at all.

CHAPTER FIFTEEN

Down in the Dumps

We went into the Salerno campaign in something of a holiday mood, despite the ninety gallon long-range tanks.

Italy's surrender on the eve of the landings gave rise to a widespread feeling of optimism. We saw ourselves living it up in Rome before the month of September was out and reaching the Alps by winter. This confidence was reinforced by the spectacle of the Eighth Army's rapid advance across the instep of the Italian peninsula and up the Adriatic Coast, where there was little initial resistance. Our optimism was ill-founded. We did not know, of course, that the Germans, so far from retiring, were pouring fresh divisions down through the Alps and rushing them southwards to join battle with the allied invaders. But within a couple of days our cheerful expectations were already clouded by doubts. It was evident that the Fifth Army was in serious trouble on the Salerno beaches. The bomb-line marked up hourly on the big map in our operations tent first bulged a little towards the interior, then crept ominously back towards the beaches. Beneath us, as we patrolled, we saw continuously the explosive signs of fierce fighting.

Enemy air reaction was sporadic. Bombers and fighter-bombers darted in and out to attack the mass of shipping standing out in the bay or to strafe and bomb the beaches. On 11 September the American battle-cruiser USS Savannah was critically damaged by a new German weapon – a radio-controlled 'flying bomb' which was launched from a distance and guided electronically onto its target. Five days later the mighty British battleship HMS Warspite suffered the same fate. I saw this great ship, surrounded by a huddle of destroyers which had to be withdrawn from support of the landings, creeping wounded through the Messina Straits on its way to Malta, where it remained disabled for many months. This sensational revelation of a successful secret weapon probably did not have much real effect on the conduct of the Salerno campaign as a whole; but its psychological effect was considerable, coming at a time when the whole enterprise, launched with such confidence, was in jeopardy.

Because of the distance separating the beachhead from the fighter bases in Sicily, it was impossible to maintain over the hard-pressed allied soldiers the weight of air support desirable both for their protection and for their morale. The original plan had envisaged the establishment of fighter squadrons on the beachhead within forty-eight hours of the landings. A couple of runways were quickly prepared by the Royal Engineer unit responsible for airfield construction, but a week after D-day the enemy was still within a few hundred yards of them and we were still operating from Sicily, one hundred and seventy-five miles away.

At this stage General Hawkins decided on drastic action. A great little man 'Shorty' Hawkins turned out to be, but at that time we hardly knew him. We saw a figure who might have been taken as a model for a *New Yorker* cartoon of an American senior officer. Short of stature, as the nickname implied, Hawkins had a lined and leathery face and a slow-speaking, gravelly voice. At all times, except when actually eating or drinking, he held a well-chewed cigar between his teeth. He now came to us and proposed a course of action which, according to the orthodox teachings of air warfare, could only end in disaster. He said that he considered it vital that fighter protection on the spot should be provided for the hard-pressed soldiers on the beachhead and he therefore wanted us to take some squadrons into one of the runways which had been prepared for us. When we pointed out diffidently that the runway in question was within easy range of enemy artillery and that we might therefore expect to lose many aircraft on the ground as a result of shelling, Hawkins merely said that this was a risk he was prepared to take.

With considerable trepidation we prepared to carry out his orders. Our fears were reinforced by the experience suffered by 93 Squadron, when they were ordered to land on the beachhead to refuel between patrols. They were being led by their CO, Ken Macdonald, who had been one of my most stalwart pilots in 56 Squadron. As the Spitfires went in to land, wheels and flaps down, they were met by a hail of fire, not from the enemy, but from our own troops. Ken prudently withdrew and asked ground control to intervene urgently with the army. Having been assured that this had been done, he made a second approach. Once again he was met by a barrage of light flak. The tail of his Spitfire was hit and severely damaged. The plane span and crashed into the ground.

And so my old friend, who, it may be remembered, had come from his family home in Brazil to fight for the old country, lies today in the Salerno War Cemetery.

In spite of this set-back, Gilroy gave orders for three squadrons to move onto the beachhead that evening. He himself took the first one in and I followed with the second. The runway was parallel with the beach and about a mile inland. Hard alongside it, on the seaward side, ran a white dusty road; and to the seaward again the road was bordered by a strip of olive trees. As I came down to land, motoring in slowly and carefully to ensure a touch-down at the very beginning of the short strip, I was nearly frightened out of my wits by a series of explosions which sounded loud above the noise of my engine and were accompanied by many flashes from the olive grove.

In spite of this unexpected distraction I landed safely and taxied in to a dispersal point between the runway and the road. As soon as I stepped onto the ground my ear drums were split by another tremendous blast from behind me. When I had picked myself up, dusted myself off and regained some dignity I asked what the hell was going on and learned that a battery of twenty-five pounders and a battery of medium field guns were lined up in the olive trees between us and the sea and were firing almost continuously.

Before landing, I had thought that German artillery would constitute a major threat to our safety and well-being. Now it seemed to me that we were in even greater danger from our own guns. I was not given to pessimism, but I was quite unable to suppress the fear that if they persisted in blasting away directly across the landing strip while our planes were in the circuit it could only be a matter of time before a Spitfire and a shell came into the same bit of sky at the same moment, with unpleasant consequences for the pilot. I turned this thought over in my mind as I toured around in a jeep, inspecting the arrangements made by the advance ground party. Finally I suggested to 'Sheep' Gilroy, who was about to take off to return to Sicily in order to oversee the move of the last two squadrons and of Wing HQ, that I should seek out the senior artillery officer and raise the matter with him. Even Gilroy's massive calm had been disturbed somewhat by the situation in which we found ourselves and he agreed with my proposal.

I took my jeep out onto the white dusty road and drove off in search of the officer commanding the twenty-five pounders. A soldier led me up to a group of tents dug in among the olive trees and asked me to wait. An officer of the Royal Artillery emerged and I introduced myself, asking if I could speak with his CO. Eventually he produced a lieutenant colonel, dressed according to best Eighth Army style in near-white corduroys and cream-coloured shirt. Our meeting was short and sweet; and it accomplished absolutely nothing.

I asked the colonel if he would kindly arrange to stop firing when our Spitfires were landing and taking off. A look of bewilderment crossed his face.

'Stop firing, my dear fellow? What on earth do you want me to stop firing for? My orders are to fire flat out, round the clock. Terribly sorry, old man, but I can't possibly stop firing.'

I explained to him my fears that one of his shells and one of my Spitfires would eventually come into contact. He looked at me as though I was absolutely mad.

'Shoot down a Spitfire! Good God, who's ever heard of a twenty-five pounder shooting down a Spitfire! Hey!' – he called over to an officer standing nearby – 'the Wing Commander here thinks we may shoot down one of his Spitfires. Take him in and give him a drink, will you, old man.'

Still muttering in amazement at the extraordinary prospect I had envisaged, he withdrew. The interview was at an end. The guns went on firing – flat out and round the clock, as specified.

I still regard it as an extraordinary thing that at no time during the next few days were my fears justified. But a much more extraordinary thing was the fact that we suffered no casualties from German artillery. A few shells came our way, but for the most part the enemy, who were short of guns and ammunition, concentrated their fire against the allied ground forces and their supply ships off-shore.

There was, however, one battery of 88mm guns which persistently caused us both danger and annoyance. It used to fire at us when we were in the circuit, coming in to land. It is unpleasant to have the black puffs of bursting ack-ack shells mushrooming around you in any circumstances; but to have it happen when you have your wheels and flaps down and are descending onto your own airfield is undignified as well as irritating. On several occasions I took off, with one or two companions, and searched for those offensive and impertinent German guns, with the intention of giving them a taste of our own. But they must have been well-camouflaged, for we never did find them.

I believe that at no other time or place, in the European theatre of war anyway, did allied fighters operate, as we did at Salerno, in front of their own field artillery. Hawkins took a tremendous risk when he sent us in there, though no doubt he was in possession of accurate information about the enemy's shortage of artillery. I do not believe that a British commander would have taken the same risk. It was a typically American action. And it typified an American attitude in war which I greatly admired – an absolute determination to do whatever was necessary in order to gain a required objective and a readiness to accept whatever punishment might come their way in the process.

This get-up-and-go approach to war frequently landed the Americans in difficulties, particularly when it was combined with the use of inexperienced troops. It also led, on occasions and particularly in the early stages of their entry into the war, to criticism from their allies, who sometimes had to help extricate them from their difficulties. A typical instance of this arose when the Americans embarked on their daylight raids against Germany with un-escorted formations of Flying Fortresses. It seemed that the critics would be justified, as the causalites mounted to a terrifying proportion of the bomber forces involved. But the Americans kept grimly on, taking awful punishment. They modified their strategy and brought in long-range fighters. They intensified their effort. For months and months on end they slogged away, fighting a long, exhausting series of titanic battles in the hostile German skies. And at last they triumphed. That long and bloody campaign waged by the US Eighth Air Force constitutes one of the most heroic feat-of-arms of the whole war and was without doubt a major factor in the defeat of Germany.

In his more limited way, our General Shorty Hawkins was a man cast in this heroic mould. The allied armies on the beachhead needed and were entitled to constant air cover. The landing strip was there, even if it was in front of our own guns. Very well, then – get on in and have a go, and to hell with the staff college egg-heads and their precious precepts. That was his attitude. And it paid off. We were able to maintain constant patrols over the beachhead and so we intercepted or deterred nearly all the hit-and-run raids which approached and thus achieved the vital objective of keeping the skies clear above the hard-pressed ground forces as they battled their way inland. We could never have done it while still operating from Sicily, which is where, I suspect, we would have remained if we had been under British tactical command.

My admiration and affection for Hawkins were greatly enhanced as a result of an incident for which he might well have had my hide. He flew in to visit us a day or two after we had arrived on the beachhead. His aircraft became unserviceable and he asked for the loan of one of our Spitfires, explaining that he was due back at his headquarters that evening for a meeting with visiting top brass. I took him round to one of the squadrons, arranged for an aircraft to be put at his disposal and away he went.

Imagine my horror when I heard an hour or two later that he had baled out into the sea. It turned out that the plane he had been given had not been refuelled. He had run out of petrol exactly half-way between Salerno and Sicily. The air-sea rescue services in the area were provided by the RAF and luckily Hawkins was familiar with the radio procedure involved. In a calm, slow voice he called three times: 'Hawkins here – Mayday, Mayday, Mayday. Baling out now.' Then he turned the Spitfire on its back and dropped out.

The air-sea rescue men had no idea who this mysterious Hawkins might be. But they got their fix and an amphibious Walrus plane was on its way in double quick time. They had the General out of the sea in under half-an-hour and he even made it to his conference on time. Instead of descending on us in wrath and fury he sent a polite message regretting the loss of our Spitfire and expressing his unlimited admiration for the efficiency of the British air-sea rescue service.

The long, hot, dusty summer broke in the end of September. It was followed by a long, cold, muddy and generally miserable winter.

The Wing was moved, on 10 October, to Capodichino, an airfield on the outskirts of Naples, now that city's international airport. (I was interested to see, when passing through the airport in August 1987, that the old pre-war hangars which had been there on the north side of the field when we had arrived forty-three years previously were still standing, externally unchanged and apparently in use.) It might be thought that after so many months in the field, constantly under canvas except for our short stay in Malta, we could have settled down in Naples to a comparatively luxurious existence. But that is not at all how I remember it. The cold rain swept down upon an ancient and decaying city from which the veneer of twentieth-century civilization had been torn by the chaos of war. The wet and draughty streets were crowded with miserable, tattered people, begging and scavenging for food. Filth was deep in the gutters. Fuel of every kind was at a premium and electricity was more often off than on.

We settled into requisitioned buildings in and around this wretched city and made ourselves as comfortable as possible. We would, I think, have been better off in tents. Sickness and infection pervaded the squadrons.

But by comparison with the armies at the front we were indeed living in luxury. The advance of Eisenhower's armies was reduced from a trot, to a walk, to a crawl and eventually to a standstill. Mud, rivers and mountains were on the side of the German defenders, still being constantly reinforced. By mid-November General Mark Clark's Fifth Army, in the west, was composed of the American 3rd, 34th, 45th Infantry, 82nd Airborne, and 1st Armoured Divisions and the British 46th, 56th (Highland) Infantry and 7th (Desert Rats) Armoured Divisions. In the east Montgomery's Eighth Army had six Divisions – the 5th, 78th, 1st Canadian, 8th Indian, 2nd New Zealand and 1st Airborne.

That mighty force faced not only many of the best divisions in the German army but also some of the most rugged terrain in Europe. Ground had to be won literally yard-by-yard on the mountainsides and in the quagmire of the valleys. At last, early in December, this slow advance stopped altogether. The front solidified along a line through the narrow waist of Italy running from Vasto, on the Adriatic, through the high snow-covered Appenine mountains to the twin bastions of Monte Cassino and Monte Camino which barred the road to Rome and would be stained by the blood of thousands of brave men of many allied nations before the winter was over.

Our job during those autumn months was to escort the medium bombers, which battered the German defences and line of communications whenever the weather allowed, and to maintain standing patrols over the front, guarding against attacks by German planes and strafing enemy movement on the roads. I was personally feeling miserably tired and dejected throughout that period. I had to force myself to fly in the daytime. In the evenings I toured the squadron messes and tried to raise my spitits by drinking freely the assorted and probably semi-poisonous wines and spirits which were purchased locally. Small numbers of enemy aircraft used to come over regularly at night in an effort to slow up and disrupt the allied advance by bombing the massive flow of materials coming in through the port of Naples. For this reason the harbour area was crammed with a tremendous concentration of anti-aircraft guns of every description. When they loosed off they produced a barrage which was almost as terrifying to the inhabitants of Naples as it was to the enemy, having regard to the fact that everything which goes up must come down.

One night I was visiting a squadron in a house on the hillside leading down from the airfield to the harbour. There was an air raid accompanied by the customary fury and deafening noise of defensive fire and the clatter of falling metal. We went on drinking through the uproar. When peace was restored the door opened and a white-faced officer came into the room. He started to say something then turned and hurried from the room, violently sick. I imagined that he had simply taken too much bad liquor, but it turned out that his stomach had been upset by something much worse than that. Upstairs, in the bathroom, one of his fellow-officers had suffered a freakish and very messy demise as a result of the barrage. He had been killed by a light ack-ack shell, presumably fired on a rather flat trajectory from somewhere lower down the hillside.

Towards the end of November 'Sheep' Gilroy was posted home. His place was taken by Wing Commander W.G.G. Duncan Smith, the wing leader of 244 Wing, who was accordingly promoted to the acting rank of group captain. I knew 'Smithie' well. He had distinguished himself as a fighter pilot at home and I had seen much of him throughout the past summer. Often, during the fighting over Sicily, our Wings had been in action at the same time and we had helped each other out wherever and whenever possible. Often, too, we had visited each other's landing fields to compare notes or to have a party. I was delighted by his posting and happy indeed to serve as his second-in-command.

Unfortunately our time together in the Wing was short. I began to feel really ill and constantly to run a temperature. I struggled along, but had to spend some time in my bed. Smithie told me that he felt forced to tell Broadhurst that in his opinion I was very much overdue for a rest from operations. With that I took to my bed again and lay there feeling utterly miserable, shivering with fever and unable to eat. The doctor said I had malaria, though it seemed a curious time of year to get it. I got up for a day or two in mid-December and then collapsed again, worse than before. I was carted off to hospital, where a mild attack of jaundice was diagnozed, and stayed there until the new year.

My place as wing leader was taken by Johnnie Loudon, the hollow-legged officer who had added such dubious distinction to our return to Coltishall after the famous American party in London, a little over twelve months earlier. I was told that I had been selected for posting to attend the next staff college course at Haifa, starting in the end of January. I had never thought much of staff colleges and similar institutions, but now I was glad to be going. I looked forward to the warm comfortable surroundings of the college at Haifa. And I felt a secret inward relief at the probability that my operational career had ended. The prospects of staying alive now seemed fairly good and in my exhausted condition I welcomed the idea of an eventual posting to a safe job on the staff.

Early in January I said goodbye to my old friends in 324 Wing. Smithie signed my log book and added the words, 'ho-ho, Cocky, staffing it at last!' and I thought that was the end of the chapter which had started three years and eight months earlier, when the first Messerschmitts curved in to shoot at me over Dunkirk. With sadness, but also with a sense of relief, I boarded a Dakota bound for Egypt.

I had the best part of three week's sick leave ahead of me and I settled down comfortably into a hotel in Cairo. Sholto Douglas, then commander-in-chief in the Middle East and Mediterranean, heard I was there and sent his car round with a note inviting me to visit him and asking if he could help over the matter of accommodation. It was an action typical of a man who was never too busy with his great and important responsibilities to interest himself in the well-being of one of his old fighter boys. But I had decided to go up to Alexandria, which I had never visited. I contacted Max Aitken, who was now a group captain commanding a mixed group with responsibility for the defence of the Delta and also for various offensive and anti-shipping operations in the Aegean. He kindly invited me to stay in his comfortable and well-staffed house outside Alexandria.

There I spent a pleasant and restful week. Max lent me a Spitfire in which I flew over on a visit to John Grandy, who was commanding the operational training unit near Ismailia, on the Suez Canal, where replacement pilots for all fighter squadrons in the Mediterranean theatre received their final training. The cold discomfort and mud of Italy faded into another far-away world as I luxuriated in the warm sunshine and enjoyed the attentions of good servants and a kind host. One evening Max suggested that I should join him for dinner at the Union Bar, one of the best and smartest restaurants in Alexandria. He would be entertaining an unknown American colonel, he explained, and would welcome my company.

Max was in a dangerously hilarious mood that evening. We drank quite a lot before dinner and by the time we arrived at the restaurant we were ready for anything. By chance all the officers of a Spitfire squadron which was shortly moving to Italy were also dining together at the Union Bar that evening. They were seated at a long table in a gallery which formed a kind of upper floor of the restaurant. Max suggested that we might enliven the situation by bombarding them with a few bread rolls. This we did, with good effect, to the amazement of our American guest. The Spitfire pilots naturally retaliated in kind and in no time a pitched battle developed. We quickly ran out of rolls and soon all kinds of food, as well as dishes and glasses, were being used as ammunition. I have a mental picture of the American colonel fighting his way out of a whole dish of spaghetti which had been dumped upon him from on high. Then all the lights went out and the uproar continued in darkness, customers screaming, waiters jabbering, tables over-turning, dishes breaking, glasses tinkling.

Clearly, that was the end of dinner, so we eventually withdrew from the devastated battle field, having joined forces with our original opponents, and moved on to a night club. There, I am sorry to report, a very similar thing happened. When Max and I finally got home we looked at each other with a wild surmise and wondered what the consequences of our night's work might be. Max was a man of decision and he immediately drew up a plan for the morning. The first thing, he announced, was to get as far away from Alexandria as possible. Accordingly we would both fly to Cairo early the following morning. He would take his Spitfire. I was to take a Magister. His driver would take all our luggage in the station-wagon. From Cairo we would drive out into the desert, where a combined army-air exercise was due to start the following evening. Max had in any case been invited to attend and I could go along, ostensibly as a member of his staff.

This movement order was carried out exactly as planned. Max's impeccable organization even included the transfer from Alexandria to Cairo of an exceedingly beautiful Egyptian girl, who seemed to be a more or less permanent ornament of his house at Alexandria. She duly appeared for lunch with us at Shepherd's. That evening we reached the Exercise HQ in the desert.

The following day, while I was hanging around pretending to take an interest in what was going on, an orderly sought me out and handed me a signal. It told me that, as a result of the death in a flying accident of Wing Commander Lance Wade, my posting to staff college was cancelled and I was to report as soon as possible to advance headquarters Desert Air Force at Vasto, in Italy, for duties on Air Vice Marshal Broadhurst's staff. I looked out across the quiet desert and contemplated the collapse of my dreams of a cushy time at Haifa. I knew that by joining Harry Broadhurst's staff at Vasto I would become enmeshed in the real war again. I told Max the news, collected my few belongings and set off for Cairo, to find a Dakota heading back whence I had come less than a fortnight before.

I knew what the job would be. Lance Wade, an American citizen who had joined the RAF early in the war and had been one of the top-scoring fighter pilots in the Mediterranean, had recently joined Broadhurst's staff in the role of GTI Fighters. I never was quite sure what the initials stood for, but the appointment was one to which Broadhurst attached a good deal of importance.

He used his GTI Fighters as his eyes and ears in all the fighter and fighter-bomber Wings in Desert Air Force. I would be provided with a Spitfire and my job would be to fly around the Wings, visiting the group captains and wing leaders and also the individual squadrons. Every evening I would report to Broadhurst in his office trailer and would discuss with him recommendations for the promotion of officers to fill flight commander and squadron commander vacancies, gallantry awards, cases of operational fatigue or, occasionally, lack of moral fibre – indeed everything connected with the fighter and fighter-bomber pilots in the command.

It was an interesting and responsible job and I was proud to have the appointment. When I reached the headquarters at Vasto, perched on the top of a windswept hill overlooking the Adriatic, Broadhurst greeted me with warmth and affection. The differences between us, which had nearly brought down my career in ruins only seven months earlier, were utterly dispelled. He had taken me into his favour, and for my part I had acquired a liking and an admiration for him which has persisted in a lasting friendship. Indeed when, in 1986, his eightieth birthday was celebrated at a lunch at the Royal Air Force Club, in a room packed with elderly retired air officers, I was detailed in advance – no doubt on his instructions – to propose the toast.

CHAPTER SIXTEEN

Close Support

At the beginning of 1944 Desert Air Force had, I believe, achieved unprecedented standards of efficiency and sophistication in the development of techniques for providing close air support for armies in the battlefield. It was a mixed force, made up of units of the Royal Air Force, the Royal Australian Air Force, the Royal Canadian Air Force, the South African Air Force and the United States Army Air Corps. The principal operational formations consisted of two Wings of Spitfire fighters (Nos 244 and 324 RAF), one Wing of Spitfire fighter-bombers (No 7 South African Air Force), three other Wings of fighter bombers variously equipped with Kittyhawks, Mustangs and Thunderbolts (Nos 239 RAF and 57 and 79 USAF), two Wings of medium bombers (Nos 232 RAF and 3 South African Air Force) and one Beaufighter night fighter squadron (No. 600).

Nearly all these units were operating from makeshift landing grounds prepared by the Eighth Army's Airfield Construction Regiment, with which we had a close and friendly relationship. Their commanding officer was an extrovert west countryman called Watson Watt. His second-in-command, Harold Keeble, was a well-known London interior decorator from whose shops in South Audley Street and, later, Walton Street my wife often bought our curtain and chair-cover materials in later years. And so Harold has, over a long period, been to me a purveyor both of airfields and of soft furnishings – the latter, paradoxically, at far greater expense to me personally than the former. The airfields which he built for us in Italy had single runways, generally about a thousand yards long and thirty yards wide, which were covered with metal known as Pierced Steel Planking – 'PSP' for short. This was an American invention and consisted of pre-fabricated and portable strips of metal which could be interlocked to form a runway or taxi- track of any length and width. It was superior to the British equivalent, which was more like wire mesh and tended to form wrinkles and ridges. In the summer, when the ground was dry and firm, we usually dispensed with the PSP.

Operations were planned jointly by the staff of Desert Air Force and their opposite numbers in the Eighth Army. The advanced headquarters of the two organizations, housed in trailers and tents, were invariably sited side-by-side, a few miles behind the line. Every evening, at about six o'clock, air and army staff officers met in the office trailer of the senior air staff officer, who at that time was Air Commodore Tom Pike – the man we had nick-named 'Killer' Pike when he was commanding a night-fighter squadron at Tangmere in the spring and summer of 1941. (He was to become chief of the air staff in the mid-fifties.) He would be joined on the RAF side by the group captain operations, the wing commander intelligence and any other officers who might be specially summoned to attend. On the Army side were the brigadier general staff, the G1 intelligence, the G1 operations and the G1 air.

The brigadier would outline the army situation and give details of his immediate intentions. He would say if there was any particular area in which close support help would be welcome and describe the type of targets likely to be involved. Following this conference, instructions would be sent to MORU (Mobile Operations Room Unit, performing much the same function as a Sector Operations Room in Fighter Command, but as its name implies, set up in such a way as to be capable of moving with the battle). There the various tasks would be sorted out and allocated to Wings. Every Wing had its own small staff of army officers – usually Eighth or Fifth army officers of company commander level, resting from frontline duty – who were responsible for giving pilots the fullest possible briefing about everything relating to their close-support tasks.

During the Eighth Army's crossing of the River Sangro a new refinement in close support had been devised by Desert Air Force. It was given the code-name 'Rover David', after the Christian name of its inventor, Group Captain David Heysham, a South African officer in the RAF who had served with distinction as a fighter-bomber pilot and who was at that time group captain operations at Desert Air Force.

The Rover principle was simple: RAF officers went out into the battle area in armoured cars fitted with radio links which connected them with the command posts of the forward fighting units. They also had VHF radio which provided two-way speech communication with fighter-bomber pilots. Whenever there was activity along the frontline, successive sections of fighter-bombers were despatched to patrol what came to be called the 'cab rank' above Rover Control. The pilots carried very large scale maps of the area, overprinted with numbered and lettered grids. They also carried, whenever available, aerial photographs gridded in a similar way. Arriving at the 'cab rank' the section leader would immediately establish radio communication with Rover Control and things might go something like this:

'Hello, Rover David, Rusty leader calling. I am with you. Do you hear me? Over.'

'Hello, Rusty leader. Receiving you loud and clear. I have some business for you. Are you ready? Over.'

'All set, Rover David.'

'Your target is on map A, square D/8. Is that understood?'

'Understand map A, square D/8. Stand by. I'll call you.'

The formation leader would then pull out the appropriate map – usually stuck down between his trouser-leg and flying boot – and identify the piece of ground corresponding to the grid reference he had been given. This was a performance which called not only for a high degree of map-reading skill, but also for a certain amount of agility and neatness. The formation leader might have had four or six maps and photographs stuck in his boot. It was a delicate task to sort these out in his cramped cockpit, select the right one, fold it to find the right square, identify the landmarks which would lead him to it and at the same time fly his aircraft steadily so that he could be followed without difficulty by the rest of the section. He would be flying at between eight and ten thousand feet, cloud permitting, and would want to take care to remain, if practicable, just his own side of the line in order to avoid, as long as possible, coming under anti-aircraft fire.

Having identified the square referred to by Rover David the formation leader would report:

'Hello, Rover David, I have square D/8. Over.'

'OK Rusty leader. Do you see a track crossing the square approximately from south-east to north-west?'

'I have that, Rover David.'

'OK. Do you see a "Z" bend in the track, with a farm house just to the north of it?'

'Yes, I see that, Rover David. Over.'

'There is an enemy mortar position just behind the farm. That is your target. Over.'

'OK, Rover David; thank you and out.'

The formation leader then had, first, to make quite sure of the target, if necessary going down alone to take a closer look at the area, and then to identify it to the other pilots in the formation, using the radio. When he was quite satisfied he would make his own attack and the rest would follow, one after the other. The leader would use his discretion, according to the intensity of the flak and the nature of the target, in deciding whether to follow up the bombing attack with a strafing run.

This 'Rover' technique was tremendously successful. It not only achieved very much more effective tangible results than the old system, when all targets had to be selected before the aircraft left the ground; it was also a wonderful thing for the morale of the soldiers fighting on the ground. A company commander whose men were pinned down and suffering casualties from some sudden and unexpected enemy fire might call for help and see planes plunging down to their rescue within five or ten minutes. The bonds between Desert Air Force and Eighth Army were thus drawn still tighter, so that the fighting men of both felt part of one combined force.

The outstanding practitioners of fighter-bombing in Italy at that time was undoubtedly 239 Wing, which operated from a landing strip alongside the beach a few miles south of our headquarters at Vasto. The six squadrons of 239 Wing were all veteran units of the desert campaigns. They had a reputation for toughness both in the air and on the ground which was unequalled within the force. This toughness was enhanced by the mixed nature of the Wing, with its two Australian squadrons and men of many nationalities serving alongside the British in the RAF squadrons.

The commanding officer of the Wing was Colonel Laurie Wilmot, of the South African Air Force. His second-in-command and Wing leader was Wing Commander Brian Eaton, of the Royal Australian Air Force. In character and appearance these two men were well fitted for their jobs. Wilmot was one of the most rugged men I ever met. He was a little over six feet tall and had a body like a huge hard rubber barrel. I have no doubt that if he had been put into the ring with Joe Louis at that time he would have survived several rounds. He had the power and the authority to control any body of men and he was also a fearless pilot.

Eaton, though a very small man, was just as tough and just as brave. He was able to command instant respect and obedience from the most unruly subordinate. Between them, he and Wilmot developed the destructive capacity of their squadrons to an extraordinary extent. Their planes were American Kittyhawks and Mustangs. Originally they had carried maximum bomb loads of five hundred pounds and a thousand pounds respectively. To Wilmot and Eaton this seemed somewhat unsatisfactory. Acting entirely on their own initiative – and that was just as well, for otherwise they would still be waiting for permission to go ahead – they carried out various tests and eventually doubled the maximum load of both planes, though they sensibly limited the use of these new and formidable loads to themselves and a few other experienced pilots.

Early in March I heard to my delight that I had been awarded the DSO. A few days later I was sitting one evening with Broadhurst in his trailer – as I did almost every evening. We were drinking a glass of whisky together and talking generally, having disposed of the day's detailed business. He asked me how soon I would feel like going back to operations. It was the question which I had been both dreading and hoping for. The old struggle was raging within me – the struggle between the knowledge that I should fight on and the desire to call it a day and stay alive. However, I said that I would go back any time he asked me to.

He then took my breath away by telling me that he was considering appointing me to command 239 Wing, as Wilmot was wanted by the South African authorities for other duties. It was a prospect both wonderful and terrifying. I was immensely flattered and immensely alarmed. Flattered, because 239 was the biggest Wing in the whole Mediterranean and considered by everyone to be unmatched in its own sphere; flattered also at the very prospect of being given a group captain's command, for such an appointment at the age of twenty-three was quite unprecedented. But this glowing satisfaction was almost offset by doubts as to my adequacy. Could I possibly assert authority over that rough, tough bunch of squadrons? I was genuinely uncertain. Furthermore I would have to learn what was to me an entirely new art – the art of fighter-bombing. Of course it was understood that there was no obligation for a group captain to do much flying. But I knew that I could not command a Wing or win the respect of my subordinates unless I regularly led formations in action. I not only knew nothing about dropping bombs, except theoretically; I was also inwardly alarmed at the prospect of having to do so day after day. In air fighting it was man-to-man and machine-to-machine; in dive-bombing you were a clay pigeon. You just had to dive down into the flak and hope for the best.

Broadie told me to mention his proposal to no one. The change would not be made for some weeks, in any case. But almost immediately afterwards the news came through that he himself was leaving. He was posted to command the air forces assembling in England to support the 21st Army Group in the invasion of Northern France.

His place as air officer commanding desert air force was taken by Air Vice Marshal William Dickson. For him the appointment must have involved much bitter personal disappointment and sadness, for he had originally been selected to command the forces now placed under Broadhurst. Like many another first class officer, both in the RAF and the army, who had been given plum jobs for the all-important Normandy invasion, he now found himself superseded at the last minute by someone from the Mediterranean theatre. That, of course, was Eisenhower's and Montgomery's doing – particularly Montgomery's. He absolutely insisted on having with him people whom he knew and people with experience of the job in hand. I think myself that he was right.

But Dickson was not only one of the most capable, he was also one of the most unselfish and delightful commanders under whom I ever served. His disappointment and dejection hardly showed and soon they were dispersed altogether, for he quickly realized that he had acquired command of a force which was probably unmatched in its own particular role and which had a fine spirit. His affection and enthusiasm for Desert Air Force became evident and obvious within a short time of his arrival. I think that before long he could honestly have said that he would not have wished to be anywhere else.

His arrival did, however, affect the plans for my personal future. Dickson naturally delayed making any major changes until he knew the people and the situations involved. Broadhurst's plan to give me command of 239 Wing had, typically, been an entirely personal selection. I was not surprised when the job went to Brian Eaton – indeed I recognized in my heart of hearts that it was the right choice.

However, there was another factor involved. I had made my own personal decision, which was that I ought to return to operations; but when, around the middle of April, I started pressing Dickson to allow me to do so he expressed some hesitation about it. He always treated me in a rather avuncular way and he suggested that I had perhaps already done enough and should be content to remain on staff duties. This opposition caused the aggressive side of my schizophrenic nature to come into the ascendant and I protested most strongly and frequently against it. I was subsequently told that Tom Pike had advised Dickson that I should be considered operationally unfit for a long time to come. I always had the greatest admiration and liking for Pike and I have no doubt that his advice was given in my best interests. But I am happy to say that in this instance he did not get his way.

Early in May, Brian Kingcome, who commanded 244 Wing, told me that his Wing Leader was in need of a rest. This was none other than my old friend Stan Turner, who had led 145 squadron in Bader's Wing at Tangmere. Kingcome, a fine fighter pilot and a legendary character, kindly suggested that it would be a good idea if I joined him in Stan's place. I agreed, I admired Kingcome and enjoyed his company. Under his command and leadership, 244 Wing had, in the victorious campaign leading from Alamein to Tunis and then again over Sicily and Southern Italy, achieved outstanding success and enhanced an already glorious reputation.

Between us, Kingcome and I managed to persuade Dickson to let me go and I simply swapped jobs with Stan Turner, who had a few days earlier been awarded the DSO. My posting came at the end of May, at the conclusion of the battle for Rome, in which 244 Wing had been heavily involved. During that battle the Germans scraped together a considerable force of planes and there was a brief resumption of fierce air fighting. 244 Wing maintained its record, destroying twenty-three enemy planes between 13 and 31 May, probably destroying three others and damaging twenty. In the course of this flurry of fighting the Wing chalked up its four hundredth victory. That achievement, combined with Stan's DSO, called for celebration, and I timed my arrival neatly. The Wing operations record book contains the following entry, describing the events of 31 May: 'It (the party) was held in an abandoned farm house, where five bars were set up, each presided over by a squadron. . . . Each vied to outdo the other in the potency of its drink, so that after a sip of some potions one felt the only thing to do was to call out the fire tender. . . . Wing Commander Dundas arrived on posting – just in time for the party, naturally.'

Naturally.

When I joined 244 Wing it was operating from a strip near Venafro, a small town at the foot of the western slopes of the Appenines midway between Naples and Rome. Seventeen days later, after a brief stay at Littorio, a permanent, pre-war airfield just north of Rome, we had moved forward on the heels of the advancing army to a new runway prepared for us near Lake Bracciano.

There was no resumption of the air fighting which had flared up during May. Our attention was directed entirely to ground targets. We flew out in sections of four or six aircraft to look for enemy transport moving on the roads and when we found it we dived down to strafe with our two cannons and four machine guns. It was not a very profitable or efficient way in which to make use of a whole Wing of Spitfires. There was a limit to the damage which could be done with guns alone against ground targets but, of course, the risks were just as great as if we had been using more powerful weapons. Indeed they were greater, for you had to get down very low in order to inflict damage. And even then some targets were impervious to mere gunfire. On 7 June we had nearly lost Squadron Leader Nevil Duke, the CO of 92 Squadron and one of the finest fighter pilots in the RAF, when he was hit by ground fire and forced to bale out behind the lines. Luckily he was rescued by Italian patriots. On 15 and 16 June four aircraft and pilots were lost.

In these circumstances, and in the light of the intelligence information available to air headquarters, which showed that the enemy fighter forces had been reduced to token numbers, it was not surprising when a load of bomb racks was dumped on our doorstep on 20 June and we received the order to prepare to make use of them without delay. There thus began for me what was, I think, the most wearing and certainly the most frightening period of the whole war, ending only when the Germans finally surrendered ten-and-a-half months later. But the first thing was to learn how to drop our bombs with reasonable accuracy. For this purpose we were allowed one week.

There were no bomb ranges available, so we practised by dropping fluorescent markers into the sea and aiming our bombs at these. Spitfires used for bombing were fitted with no special devices. We had to use the ring-and-bead reflector sight intended for air fighting. But it was extraordinary how accurate a good pilot could become with experimentation and experience.

We each had to develop and try to perfect our own technique for achieving accuracy. In due course I found that if I flew so that the target passed under my wing just outside the cannon mounting, then held my course until it reappeared aft of my wing, I would be in about the right position to begin my dive. The target would thus be a little to one side and very slightly behind. It was then necessary to turn the Spitfire over onto its back and let the nose drop through the vertical, using ailerons and elevators to position the red bead of the reflector sight on the target and hold it there. The angle of dive would be about twenty degrees off the vertical and this would be held from the starting height of about eight thousand feet to something under two thousand feet. At this point I would decrease the angle slightly to bring the bead ahead of the target, at the same time counting 'One-and-two-and-three', then press the button. No doubt the whole procedure sounds thoroughly Heath Robinson, but it worked. In due course I reached the stage where I was most dissatisfied if my bomb burst more than fifty yards from target – and a five hundred pound bomb exploding only fifty yards away can be rather more than an irritant. Usually I succeeded in doing much better than that.

We were given our first operational bombing target on 26 June, when Brian Kingcome and I each led a Flight of 417 Canadian Squadron to bomb a road junction. The target was no doubt chosen with a view to breaking us in gently. It was well behind the lines, so there was no chance of bombing our own troops, but in a quiet area. We encountered no flak at the target and saw no enemy movements. We dropped our bombs with indifferent accuracy and returned to base feeling rather let down.

It was a feeling which did not last long. Gradually we were brought more into the thick of things and given tougher targets. The battle on the ground was still fluid, with the enemy falling back to dig into and fortify the 'Gothic Line', a series of linked strong points stretching across rugged mountainous country from the west coast, north of Pisa, to Pesaro on the Adriatic. Kesselring, our old Battle of Britain opponent who now commanded all German forces in Italy, had an army of 'Todt slaves' at work, preparing the line. His purpose was to delay the allied advance as long as possible, while the defences were completed. And our task was to make life as difficult for him as possible by interrupting the movement of his troops and supplies on the roads wherever we could find it.

This work went forward with great intensity after we moved to Perugia on 3 July. It was good preparation for the very much more exacting form of fighter-bombing which was to follow when the armies were locked together again in battle and we became engaged in true close support work. Our interdiction sorties gave us an opportunity to perfect our map reading technique, which had to be accurate to a degree not normally required in a fighter pilot. They were not so demanding in that respect as the very close support work which would soon follow, but we gained experience which stood us in good stead a little later. Among other things we began to be wise in the ways of the enemy flak. There was little to be done about that menace once you were committed to attacking a target, but we developed an intense interest in intelligence reports about flak concentrations and acquired a certain instinct for approaching and leaving a target by routes which might be expected to be flak-free.

I received my first serious hit from flak on 10 July while leading a formation on an interdiction mission near Arezzo. There was a thud as an 88mm shell burst just beside me and several fragments hit my plane. The glycol tank was punctured and once again I found myself sitting in a cockpit full of thick, hot white smoke. I immediately jettisoned my bomb and went into a diving turn, at the same time reaching up with my free hand to pull the little rubber knob which would release my cockpit canopy. As I did so, I received a stunning blow across the head which knocked me practically senseless, so that for some seconds I hardly knew what I was doing. As I struggled back to full consciousness I became aware of the fact that my cockpit canopy, instead of flying off upwards and outwards, had crashed down upon me. The metal strip along its front edge had slashed across my face with all the force of a 240 m.p.h. wind behind it.

However, the smoke was now pouring out of the cockpit and, ducking my head down, I pushed upwards with my forearm and elbow to force the canopy up and away. Dazed, battered and thoroughly frightened I took stock of my situation and my position. I was down to about five thousand feet and my engine temperature was mounting fast. Below me was rugged, hilly country on which a forced landing could only end in disaster. I set course for Castiglione, the nearest friendly landing ground and just clear of the hills. But I did not expect to make it. I thought I would almost certainly have to take to my parachute – an imminent prospect which I dreaded.

The miles slid past beneath me. The needle on the temperature gauge circled slowly but perceptively towards the danger mark. It was going to be a devilish close thing, but I began to think I might make it as I saw the waters of Lake Castiglione show up through the haze ahead. The needle passed one hundred and ten degrees . . . one hundred and fifteen . . . one hundred and twenty! Three more minutes perhaps and about eight more miles. I clung to all the height I could hold and flew gingerly on. Luckily the runway was in line, dead ahead. At last I realized I had the height to make it, with or without engine. One hundred and twenty-five degrees and going quickly off the clock now. I slammed down the wheels, cut the switches, turned off the petrol and glided down. My propeller locked solid as I finished my landing run. I sat still and listened to the lovely silence. Then I put up my hand to unclip my oxygen mask and found that my face was sticky with blood.

It was lucky for me that I had instinctively pulled my goggles over my eyes as soon as I had been hit and the cockpit had filled with smoke. The strong metal frame of the goggles, together with the thick rubber of my oxygen mask, had probably saved me from being knocked out completely when the front of the hood smashed down onto my face. The impact had cut through the lower part of the frame on my goggles and had sliced the oxygen mask. I had a deep gash running from the corner of my right eye down across my cheek.

I ordered the ambulance driver to take me first to the officers' mess tent and there I got a large glass of brandy with which I re-embarked in the ambulance *en route* for the doctor. I observed that, though it was only five o'clock in the afternoon, there was something of a party going on in the mess and a South African major, whom I knew only slightly, was in a state of advanced exhilaration. A few minutes later I heard a Harvard trainer – used in many squadrons as a 'communications' aircraft – taking off. As I was lying on the doctor's couch, anxiously watching the needle which was held poised to sew up my cheek, the peace of the evening was broken by the noise of the Harvard diving immediately above us. As the needle descended, so also did the Harvard, the shrill note of its engine getting ominously close.

I can hardly blame the doctor for the slip of the hand at the critical moment, which resulted in his needle digging in over-deep. For the Harvard hit the ground in vertical dive with a terrifying crash and explosion not much more than a hundred yards from where I lay. Thus, uselessly and tragically, ended the life of the South African major.

In spite of this distraction the doctor made a good job of stitching me up. Over the years the mark has disappeared, though an entry in the Wing diary records that I returned that evening 'bearing as handsome a duelling scar as ever was seen'.

Soon after this episode His Majesty King George VI visited Perugia, where he landed before making a tour of the Italian Front. By good fortune the honour of meeting him fell to me, for I was in charge at the time, while Brian Kingcome was away in England on leave. General Oliver Leese, who had taken over the Eighth Army when Montgomery left, and Air Vice Marshal Dickson arrived on the scene about half-an-hour before the King's plane was due and we drove round the airfield together in order to ensure that everything was in order.

After the king had landed we all got into Leese' open Humber staff car, still painted in the yellow and clay-coloured camouflage which many men who had fought in the Desert still stuck to, more out of sentiment than for any other good reason. The king sat in the back with Dickson; Leese drove and I sat beside him in the front. We drove slowly round the perimeter track, where the officers and men of each squadron stood in front of their Spitfires and cheered their king's passing. Afterwards I took my distinguished guests to my caravan, where an awning had been set up and my batman had laid out the best liquid and solid refreshment we could offer.

Dickson stayed behind after the King had left with Leese. We sat and talked and had another drink. After a bit of humming and ha-hing, Dickson told me that he was rather unhappy about one aspect of the King's programme. He explained that His Majesty was going to visit an Artillery Observation Post for a close-up view of the battlefield. From there he would be driven back to one of the tiny landing strips used by the little Auster aircraft which served as artillery spotter planes. These aircraft were flown by army pilots of the Royal Artillery and the intention was that one of them should then fly the King back to army headquarters.

Dickson told me that he felt that any flying which the King might do in that area was his responsibility. He said that he was considering the advisability of taking his own Storch – a captured German communications plane capable of landing in exceptionally small areas – to the forward landing strip and offering to fly the King back personally. He asked my opinion. I had no strong views on the subject, but encouraged him to carry out his plan, feeling that this was the diplomatic line for me to take.

The outcome was comical but extremely unfortunate for poor Dickson. He duly set out on his well-meant mission and arrived over the landing strip in his Storch just as the King, the army commander and their respective retinues drew up in a convoy of jeeps. The appearance of the Storch was a surprise to one and all, but it was instantly recognized as Dickson's plane and all the top brass lined up beside the little strip to watch his landing. A Storch could touch down and draw to a standstill within a distance of about forty yards and Dickson made his approach in such a way as to land immediately in front of the red-tabbed spectators. He judged things to perfection, in so far as the point of touchdown was concerned. But most unfortunately he somehow contrived to land with his parking brake locked on. The consequence was that the moment his wheels touched the ground the Storch up-ended and came to a standstill in a cloud of dust with its wheels pointing up to the heavens and the air officer commanding Desert Air Force hanging upside-down in the cockpit.

Luckily he was unhurt. But it came a little hard for Dickson, after he had been helped out by a group of surprised but sympathetic generals who wanted to know what on earth he was doing there anyway, to explain that he had just dropped in on the off chance that the King might prefer to fly with him rather than an army pilot.

While the Wing was at Perugia we made some attempts – mostly abortive – to organize social contact with the local population. The Wing log, which at that time was being written in racy narrative style, describes one such effort as follows:

> An early visitor was the priest of a neighbouring parish, who wrung the withers of the senior adminstrative officer with his sad stories about the war damage sustained by his church. He pointed to a corner of the airfield, where there was a pile of rubble and a collection of assorted equipment and bric-a-brac left behind by the Germans, and asked permission to cart some of it away to use in the rebuilding of the church. Inspired either by the milk of human kindness, or possibly by the demon vino with which he was entertaining his reverend guest, our admin officer told the priest to go right ahead. Next morning there descended on the airfield an army of parishioners, equipped with every wheeled conveyance available. It soon began to look as though their purpose was to build a new church about the size of St Peter's. Every moveable object in the camp was disappearing into their carts. Our clerical visitor was bustling around at top speed urging his parishioners on to greater efforts. Immediate and energetic action was required to restore the situation and strong forces of airmen were directed to engage the marauders. The priest and his parishioners withdrew, chattering like magpies.

In Assisi we took over an hotel for the officers of the Wing. Our first effort at organizing a dance is briefly described in the Wing diary: 'Special late passes had been provided by the Town Major for some popsies of not very exceptional beauty who, to varying degrees of disappointment, turned out to be of the 'haut' rather than of the 'demi' monde. Their scruples did not deter them however from toting away as much of the buffet as they could cram into their handbags.'

I can state with confidence that during the few weeks we spent at Perugia some of these ladies were persuaded to lower their status somewhat and to investigate the pleasures of the demi-monde. But these social exercises were short-lived. The battle for the Gothic Line was about to begin. And in preparation for it the whole Eighth Army and Desert Air Force was concentrated on the Adriatic Coast, where the fight was to be fiercest. On 26 August we flew away from Perugia and landed on a strip prepared for us at Loretto, a few miles north of the port of Ancona, on the Adriatic coast.

The map of Italy shows very clearly the problem confronting the Army Commander at that time. In order to break into the Po Valley it was necessary first to squeeze through the mouth of a funnel formed by the sea on the one side and the mountains on the other. Between Pesaro and Rimini the coastal road runs up a narrow strip of land, with the sea a mile or two to the east and the steep mountains reaching down almost as far as the road on the west side. Here, on 25 August, the Eighth Army threw itself upon the heavily fortified Gothic Line. Here we began in earnest the operation of a close support fighter-bomber role.

Day-in day-out, from misty dawn, through the fierce heat of midday, in the comparative cool of the evening, we took off, climbed out over the line a few miles to the north of our landing strip and plunged down into the fire of the battle. The flak was intense and terrifying in those steep valleys. It could not be evaded. It poured up at us in every known form – big black puffs powdering the sky around us as we ran in to the target, thousands of little white puffs lower down as we dived, streaks of tracer hissing across the valleys as we dropped our bombs and clawed upwards again, throttles wide open, seeking the comparative safety of altitude.

I watched the faces of the pilots in the briefing tent as the army liaison officers and operations officers explained the details of their tasks before take-off. Every man knew that within half-an-hour he would be facing the flak, quite unable to defend himself against it. When you said 'Good luck' to a pilot going off on one of these missions that was literally what you meant. Good luck. That was your hope. That was what you clung to. With good luck you would get through the barrage unscathed. With only a bit of luck you would sustain a hit, but would still be able to fly home. With bad luck you would be forced to bale out or to crash land. With no luck at all you would be killed. It was not surprising that I often saw fear in the eyes of those young men as they listened to the briefings. It is not surprising that from time to time someone would find it almost impossible to control his fear.

It was my job to ensure that fear was held within restraint. If it took hold it would quickly spread through every squadron of the Wing. And yet there was no one who felt more afraid than I, so the job was a hard one.

There was a road bridge north of Rimini, a vitally important link in the enemy supply line, which we were ordered to destroy and keep out of action. It was a target we all hated, because around it the flak was particularly intense. One morning we were ordered to put a whole squadron onto it.

As I had to decide which squadron would do the job, it seemed only fair that I myself should lead it. We flew out through a clear sky, no cloud to hide our coming; I followed the coast, flying a few miles out to sea. North of Rimini I turned and led the way in. The black puffs burst all around us before we had even crossed the coast, with the target still four or five miles ahead. The temptation to swerve away was almost overpowering. I felt naked and exposed and was sure that I was going to be hit. The target passed under my wing and I rolled over into a dive. Down through the black bursts, down headlong into the carpet of white, where the 40mm shells came up in their myriads to meet me, down further into the streaking tracer of machine gun fire. I dropped my bomb and kept on down – safer on the deck than climbing up again – and used the R/T to tell the others to do the same, everyone to make his own way back across the line. Just after I had transmitted there was a thudding explosion and my Spitfire juddered. Bloody hell! A great hole had been ripped in my port wing, half-way between cockpit and wing tip. But she kept on flying and I held my course and speed, gaining height as soon as I had crossed the line.

I flew home without difficulty. My wheels and flaps came down all right so I went in for a normal landing. But one of my tyres had been punctured. As soon as the weight came down on it the plane slewed round and I could not hold it. The undercarriage leg collapsed and I ended up with one wing-tip on the ground.

I went to the caravan where Brian Kingcome and I had our living quarters. He was entertaining Duncan Smith, who had flown over to see us. They both treated my adventure as a huge joke – quite rightly, too. But for once I was not feeling jokey. I told them to go to hell and lay down on my bunk and thought, 'Oh Christ, Oh Christ, I can't go on like this.'

But of course I had to go on like that. There was no acceptable alternative. The squadron pilots knew well enough what were my feelings about men who held back from the task. If a squadron commander was doubtful about the 'moral fibre' of one of his pilots he would discuss the matter informally with me before taking any other action. I would then arrange to fly with the squadron on an operation and take the suspect as my number two. Thus I had the opportunity of judging for myself whether he was pressing home his attack under fire. It was a cause for some pride and satisfaction that, despite the excessive strain imposed by the nature of our work, only a very small number had to be dealt with for 'lack of moral fibre'.

There was one man who, I firmly believed at the time, actually tried to shoot me down because he realized that I suspected his courage. He had come to the Wing as a flight lieutenant, though not as a flight commander, for he had little operational experience. His squadron commander thought little of him and eventually told me that he believed him to be lacking in courage. Accordingly I took him with me on a sortie and ordered him to follow me on a strafing run after we had dropped our bombs. There was heavy flak around the target. I held my dive until I suddenly became aware that tracer was going past – and uncomfortably close – from behind and above as well as from below. I looked quickly in my mirror. And there was my number two, sitting about two hundred yards away directly on my tail, blazing away for all he was worth. Breaking violently, I ordered him to form up beside me and we flew home.

Back in the pilot's tent I asked him what the hell he thought he had been doing. He said that he had just begun his strafe when I broke away and ordered him to reform. Of course I could not accuse him openly of trying to do away with me. But I did have him slung out in quick order. And I think he knew why. To this day I believe that my life was probably saved on that occasion by the lucky chance that he was flying one of the new batch of planes only delivered a few days earlier. These new planes were the only Spitfires in the Wing which carried .5mm machine guns, instead of the usual .303mm Brownings, outboard of the cannons. And the ammunition belts provided for the .5mm guns were fitted with one tracer round in every five. None of our other guns fired tracer. And but for that tell-tale tracer I would never have looked in my mirror.

Despite the constant danger and the constant casualties, morale in the Wing can never have been higher than in those weeks of battle around Rimini. A spirit of fierce pride developed throughout the squadrons as we gained proficiency in our new role. It was rewarding to hear again and again that some local objective had been captured, or some hard-pressed body of troops had been relieved, as a result of our work. Messages of congratulation and appreciation arrived regularly from the ground forces we were supporting and we came to identify ourselves with them, and with their victories and their setbacks, as never before.

By October the gallant Eighth Army had slogged its way through to the plain and was fighting its way up the long straight road which led to Bologna. But by October the rains had also come. The advance, always opposed yard by yard, slowed to a crawl. Finally our troops were halted altogether on the banks of the River Senio. It was only a narrow river and any tourist driving by might wonder how such an insignificant trickle could have held up a mighty army for so long. But it had strong and high flood banks on either side, built up to a height of twenty or thirty feet in many places. Into these banks the opposing armies dug their forward positions and there they growled at each other through the long hard winter. There were fierce bursts of local action all along the line, particularly on the flanks where, to the seaward side, the Senio flowed ill-defined through a kind of delta, and to the south-west the foothills reached down to the water-logged plain. But the general advance was halted and would not begin again until the spring.

On 11 November Brian Kingcome was posted home to the United Kingdom. A day or two later Air Vice Marshal Dickson asked me to go to his headquarters to see him one evening. I supposed that he was going to tell me who Kingcome's successor would be and when he was due to arrive. Or possibly, I thought, he wanted to tell me that I was to be withdrawn from operations and rested, though there seemed little chance of that, as it was unlikely that he would change both the CO and the second-in-command of a Wing simultaneously.

Dickson sat me down and gave me a glass of whisky. He told me that he was leaving Desert Air Force to go back to the Air Ministry. He recalled how bitterly disappointed he had been when he had been posted to us, losing his command for the second front. But now, he said, there was no force in the world that he would rather lead than Desert Air Force and he was terribly upset about having to go. Then he started talking about 244 Wing and was warm in his praise of our achievements during the recent battle. There had been one thing he was determined to do before leaving: that was to get Air Ministry approval for my appointment as commanding officer to succeed Kingcome, with the acting rank of group captain. There had been some opposition, because of my age. But he had persevered, approval had come through that day. I could put up my fourth stripe and get hold of some gold braid for my hat. And who did I want as my wing commander flying?

I was immensely proud but also shaken by the thought of the responsibilities I was to face. I think no officer in the RAF had ever been promoted to group captain at so young an age – I was then only three months past my twenty-fourth birthday. The honour was therefore something of which, I believe, I had a right to feel proud. But the responsibilities were formidable indeed. I had to answer for the operational efficiency and administration of five squadrons in the field. I had therefore about a hundred aircraft on the front line under my command and as many pilots. Including the repair and inspection unit which was attached to the Wing and was part of it for adminstrative purposes I had about two thousand men under my command in all and was entirely responsible for their efficiency and their well-being.

My doubts and fears were sharpened by the knowledge, in my heart of hearts, that I was just about played out so far as fighting was concerned. But I knew also that I could not be a chairborne CO. I must lead my Wing in the air as well as on the ground. Anyway, I had been given the job. I would have to stick it out until the end. And so, in a haze of happiness and uncertainty, I drove back to the airfield and told my astonished batman to get going with a needle and thread and turn me into a group captain.

CHAPTER SEVENTEEN

The Last Throw

In theory I should at that stage have hung up my flying boots, taken up station in the caravan which served as my office and watched the replacement wing leader get on with the conduct of flying operations.

In practice it did not work out that way – for two reasons. The first reason was that Wing Commander Pete Lovell, who was posted from MORU on 21 November to take over as wing commander flying, was posted away again only nine days later. We had to wait until the end of December for the arrival of his replacement, Wing Commander Ron Barry, DFC – A New Zealander whom I knew well. So I just carried on as before, spending most of my time supervising the Wing's operations and frequently taking part in them.

The second reason why nothing much changed, except the stripes on my tunic and the addition of some gold braid to my cap peak, lay in my frame of mind during that last wartime winter and the early spring of 1945 which took us at last to victory. Many years later I was asked by Douglas Bader to contribute a chapter to a book he was editing about the Spitfire aircraft at war. My bit was about the Spitfire's role in Italy. In describing the part we played in preparing for and then supporting the last great advance from the River Senio to the Alps, I wrote that I thought that some of us were, by that time, 'a bit dotty'.

Perhaps 'punch drunk' would have been a more precise description of my condition. I know that I was consumed by a fierce determination that 244 Wing should be pre-eminent, performing its role in supporting the Army with the utmost energy and diligence. Nor was I content with the weapons and methods provided and laid down for our use in the process. Taking a leaf from Colonel Laurie Wilmot's book, I had a double bomb-rack designed and assembled and by the end of January Ron Barry and I were both carrying two five hundred pound bombs under our Spitfires, instead of one. We also devised new tactics which enabled us to provide close support for the army in weather conditions which would previously have been regarded as non-operational. I think that at this late stage, more than at any other time in the previous five years, a sincere desire to engage the enemy had at last got the upper hand over a simple desire to stay alive.

There was, anyway, plenty of opportunity to give rein to that state of mind. I have been surprised, in reading through the Wing Log, still stored in defence ministry archives, by the level of activity maintained throughout that winter. The army's advance had been halted in October, but there was a constant call for support along the front line which followed the Senio river across the Adriatic plain, north of Rimini and south of Bologna, and then squiggled up into the mountains. We were also required, whenever the weather allowed, to prowl about behind the line, attacking transport, bombing railway communications and generally doing all we could to hamper and disrupt the movement of supplies intended for the enemy army.

Our ability to cover the enemy's long lines of communication was enhanced when, on 4 December, we were moved up closer to the front, to a landing strip made of the now-familiar PSP (Pierced Steel Planking) at a little place called Bellaria, on the coast about six miles north of Rimini. The runway, parallel with the beach and divided from it only by a narrow road and some sand dunes, was exactly a thousand yards long and twenty-five yards wide. Around it, but mainly on the inland side, the Spitfires of our five squadrons – between ninety and a hundred aircraft altogether – were dispersed to the greatest extent possible along such taxi tracks as the Royal Engineers had been able to provide in the limited space available. Furthest away from the runway was the Wing Maintenance Unit, constantly struggling to repair the Spitfires which, day in, day out, came home damaged by flak.

Our living and office quarters were sited, some in caravans, some in houses, in the village which ribboned along the coast north of the runway. There, for the first time since I had flown back to Italy from my brief visit to Egypt the previous January, I established my own quarters in a house rather than a caravan. It was on the coast road and had, I imagine, been built originally as a small holiday villa. I was joined there in due course by Ron Barry and we were looked after by Corporal Butell – a rough, tough and fiercely loyal London docker who had long been Brian Kingcome's devoted servant – my driver, Leading Aircraftsman Windsor, and a displaced Italian professor from Bologna University who had found himself on the wrong side of the lines without visible means of support and so, in return for food, shelter and some small amount of money, had taken on the more menial and domestic chores in our little household.

An indication of the high level of operational activity carried on by the Wing, despite the winter weather and the immobility of the army, is to be seen in the fact that on the very day we moved from Fano to Bellaria twenty-five separate operations were carried out, involving one hundred and fifty-five aircraft – one of which did not return. That high level was kept up right through December, the Log recording one hundred and fifty-two sorties on 10 December, one hundred and forty-five on 11 December and, on 15 and 19 December, two successful applications of a new type of attack, largely of my own invention, code-named 'Rover Piggy'. Already a variation of the Rover

David technique had been dreamed up and was used regularly when the cloud base was too low for dive-bombing. It was named 'Rover Timothy' and involved attacks against ground targets, using only cannon and machine guns – no bombs. One day in mid-December the army called urgently for our help in raiding an enemy outpost which was causing casualties and holding up an attempt to capture some ground on the Canadian Corps front. But the weather conditions seemed too bad, even for a 'Timothy' operation.

However, it became clear to me, from the insistent and rather desperate nature of the calls for help, that our intervention was badly needed. I took off to see for myself whether cloud base and visibility in the area really ruled us out. Conditions were indeed extremely murky, but the ground was flat – no hills to bump into. After landing I spoke to the Controller at MORU and asked him to suggest to the army that they should, at a predetermined moment when I would endeavour to arrive on the scene with a squadron of Spitfires, lay down a line of white smoke shells immediately in front of the area which they wanted us to strafe.

The offer was quickly taken up and I was soon airborne with, appropriately, 417 Squadron – our own Canadians. As we arrived on the scene there duly appeared on the ground a thin line of white smoke and down we came, out of the gloom, aircraft echelon starboard, cannons thudding and machine guns rattling away. We could not see much in the way of targets, but that was not important. The army had shown us exactly where the trouble was coming from. The area was small, as well as clearly defined. And twelve Spitfires, mounting between them twenty-four cannons and forty-eight machine guns, could hardly fail at least to intimidate the foe. We were later to have confirmation that we had, in fact, done a bit more than that, when a signal was received from the Royal Edmonton Regiment thanking us for our efforts and adding: 'When slit trenches along the canal were captured they were found full of defunct Huns – over thirty killed by strafing.'

I was fortunate, particularly during the time when there was no wing leader, in the quality of the men I had supporting me that winter, on the ground as well as in the air. Notable among the former was Charles Abrahams, whose job it was to supervise the 'Ops tent', a small marquee where we assembled for briefing before every mission. Charles – a lifelong friend who, after the war, was chairman of Aquascutum and was made a KCVO for his public services – was then, as he always remained, a most kindly man who felt very deeply his responsibility in relation to our dangerous job. Among the men commanding our squadrons that winter. Danny Daniels, who had been with me in Malta and Sicily, was back after a short rest, commanding 145 Squadron. He was no less notable as an aggressive fighter bomber leader than he had been when fighting Messerschmitts. He was shot down by flak on 27 December and slightly wounded, crash-landing close to the front line. But he got away with it and was back with us well before the campaign was over. Another man I knew well was 'Stimmie' Stimpson. He had been a sergeant pilot with me in 56 Squadron, back in the Duxford days, and now he commanded 601 Squadron. No. 241 Squadron, the latest addition to the Wing, was commanded by Mike le Bas, who went on to a distinguished career as an air officer and remained a good friend until his death in early 1988.

The most outstanding leader of them all was a young South African, Major John Gasson. He had joined the Wing in the Desert, as a very young Second Lieutenant and had greatly distinguished himself as an air fighter, winning the DFC. When I joined 244 as wing leader, he had just been promoted to command a flight, with the rank of captain. In due course he was promoted again, on my recommendation, to command 92 Squadron. He was one of the bravest men I ever knew, and one of the best. We celebrated his twenty-first birthday in mid-February and by coincidence we also celebrated, at the same party, the award to him of the DSO. It seems extraordinary, looking back at it, that 92 Squadron, one of the most famous in the RAF, should have been commanded in that last crucial campaign by a South African Major who had not yet come of age.

Johnnie survived to the end, though not for want of constant risk-taking. As soon as possible after the war was over he went home to study medicine. For nearly forty years he has pursued his second career as one of his country's most brilliant surgeons. We have been lifelong friends.

Soon after Johnnie's DSO was gazetted we had an excuse for yet another party. On 23 March a signal was received to say that I had been awarded a Bar to mine.

Very early in April we knew that the final attempt to break out of the winter line on the Senio and to drive the Germans back across the Po and out of Italy was soon to begin.

All the group captains commanding Wings and their wing leaders were summoned to a meeting at Desert Air Force advance headquarters and given a detailed briefing. Our AOC, Air Vice Marshal 'Pussy' Foster, was supported not only by members of his own air staff, but also by senior army officers, who explained to us in detail exactly what their plans and problems were.

The Senio was quite a small river in itself. But it had very high, very broad, very solid flood banks. The Germans were strongly entrenched and established in defensive positions all along the north bank and so that was the initial obstacle. Then, about a hundred miles north across the coastal plain, there was the broad, fast-flowing River Po. It was made clear to us that the army was relying heavily on massive air support – first, very close support in helping them to break out of the Senio line, then non-stop attacks on the enemy's movements behind the lines. The intention was that we should dish out to the German ground forces treatment similar to that experienced by British and French troops during the fall of France and the Low Countries five years earlier.

The plan was to use our air power in order, firstly, to disrupt the passage of supplies southwards to the German armies and then to make it as difficult as possible for the enemy to retreat northwards across the Po, by cutting the road and rail bridges across the river and by destroying the fleets of small boats assembled to serve as ferries. The success of this strategy was reflected in a statement made soon afterwards by the German land force commander, General von Schwerin.

'The bombing of the Po crossings finished us. We could have been successfully withdrawn, with normal rearguard action, despite the heavy pressure. But due to the destruction of the ferries and river crossings we lost all our equipment. We were no longer an army. The attacks were demoralizing. Towards the end our soldiers could not fight.'

The break-out across the Senio started on 9 April, in the middle of the day. The Wing flew 121 sorties that afternoon, all close support. We were given the honourable role of strafing a stretch of the north bank of the Senio, at one of the breakthrough points, while our own troops waited in and behind the south bank, separated from our target area by a distance equivalent only to the length of two or three cricket pitches.

For the next eighteen days we were at it day in, day out. From dawn until dusk our Spitfires were taking off from Bellaria, sometimes with preordained targets, more often, as the battle developed and the Germans fell back, on free-ranging missions to seek out and destroy enemy transport of every kind, from horse and ox-drawn vehicles, which abounded in those last months of the war, presumably because of the enemy's shortage of petrol and oil, to heavy tanks.

Very early on, when the success of the ground attack was still in the balance, I was faced with the need to make a terrible decision. Although I did not realize it at the time, it was foreshadowed by a curious and rather eerie incident, a day or two before the campaign began. I was leading a dive-bombing sortie, had just completed my own attack and was zooming up again, at the same time watching the others dive down. My eyes were on one of them, descending vertically towards the target, perhaps four or five hundred yards from me, when, in a split second of time, that Spitfire just ceased to exist.

I had never before seen anything like it. And although the memory of the incident has stayed clear in my mind right across the years, I have never been able to translate it adequately into words. At one moment there was the familiar, solid shape of a Spitfire; at the next, a flash, as though the aircraft had been vaporized and turned into a million pieces of confetti. In reporting this incident, I expressed the view that the detonator on the bomb must have sustained a direct, flukish hit by enemy anti-aircraft fire.

On 12 April Ron Barry was killed — a sad and terrible blow. The pilots who had been flying with him reported that his aircraft 'had exploded in the bomb-dive'. Their description closely matched what I had seen two or three days earlier.

Within hours, I received a visit from an officer from Desert Air Force. He told me that a batch of suspect bombs had recently been delivered. A sample taken somewhere back up the supply line had shown that a very small proportion had faulty fuses, which could lead to the bombs exploding in the dive. A number of these potentially lethal objects had been delivered to Desert Air Force, but the only ones actually sent on down to units in the field had come to 244 Wing. They were somewhere among the several hundred bombs held by our squadrons – at the time we were using about a hundred and fifty bombs every day – and, although in retrospect it seems surprising, there was no way of identifying them. There was, therefore, only one way of making absolutely certain that no more pilots would be blown to smithereens in their bomb dive. That was to suspend operations while our entire stock of bombs was put on one side and replaced.

I called the squadron commanders to a meeting in my house. We had a drink and talked it over for a few minutes. And then we reached the inevitable conclusion. We could not suspend the operations of our five squadrons at that critical moment. We must carry on.

There was no further discussion of the matter, at any level – not between me and the squadron commanders, not with anyone at Desert Air Force. I do not know whether the AOC even knew about it. If he did, he never mentioned it to me. There is no written record of it in the Wing Log. Naturally, neither the squadron commanders nor I said anything to the pilots, or indeed to anyone else. A number of people, in the squadrons and in the wing armoury, must have known that there was a holus-bolus changeover of bombs taking place, but presumably some plausible reason was given and accepted.

There were no more incidents such as I had witnessed. But for two or three days those of us who had sat in my house and had decided to carry on felt that a new dimension had been added to the always chancy business of dive-bombing.

The battle which began on the Senio river on 9 April was virtually over by the end of the month. Our own activity was at its most hectic during the middle week of that period, when, day after day, we flew more than a hundred and fifty sorties, reaching a record of a hundred and eighty-two on 19 April. Mostly we hunted our targets in small formations of four or eight aircraft. But on occasions we went out in strength. The log records that on 18 April, 'we carried out another of our celebrated Timothies; three squadrons led by the Group Captain attacked buildings south-east of Budrio very successfully.' I believe that this was the only occasion when I led a single formation of more than one squadron on a ground attack mission.

The pressure was intense – and not only on the pilots. The ground crews who maintained, refuelled and re-armed our Spitfires had no rest. The men in the Wing maintenance unit, who undertook repairs which were beyond the scope of the squadrons, were working round the clock. So also was the armament section – particularly that part of it which had to load cannon shells and machine gun bullets into belts. That section was reinforced from all sides, and my own report, written at the end of the campaign, records that 'waiters, photographers, sanitary squad, all came to the rescue.'

Between 9 and 25 April, 244 Wing succeeded in flying more sorties each day than any other Wing in Desert Air Force. The statistics are formidable. In the course of the campaign the Wing flew 4,644 operational hours, comprising 3,661 sorties; 851 tons of bomb were dropped; 345,380 rounds of 20mm cannon and 804,291 rounds of machine gun ammunition were expended. But we were receivers as well as deliverers of this deadly stuff. Anti-aircraft fire resulted in fifty-five of our aircraft being destroyed or so badly damaged as to be beyond repair; forty-one more were damaged to the extent that we could not repair them ourselves; and another thirty-one were hit, but could be put back into service by our own ground crews. So, altogether, one hundred and twenty-seven aircraft were hit. As we would have started out with between ninety and a hundred, the strike rate against us in those three weeks was well over a hundred per cent.

And what about pilots? Three or four weeks after it was all over I wrote my own comments on that subject:

'Losses of pilots were not severe. Nine were killed, or missing believed killed. Three more were missing, one of whom has since been accounted for as a POW. Three more were injured. Several baled out or crashlanded behind the lines and returned within a few hours or a few days, Flight Lieutenant Jones, DFC, of 601 Squadron, repeating this performance twice.'

I was a little surprised, reading those words forty-three years after they were written, that I should at the time have considered the losses 'not severe'. The maximum pilot strength of a squadron was eighteen men. In practice it normally averaged fifteen or sixteen, so on 9 April we probably had about eighty pilots, including Ron Barry and myself. Of those, nearly fifteen per cent were dead or missing when the shooting stopped, three weeks later. And a good many more than that, having regard to the number of Spitfires which were completely written off, must have baled out or force-landed. Not a severe rate of loss? It is an illuminating comment.

I, personally, came through that last frenetic three weeks without a scratch, either to myself or to my Spitfire. My faithful Italian servant – the displaced professor from Bologna University – had made a pathetic but heartwarming attempt to ground me at the very beginning of the affair. He was so shaken by poor Ron Barry's death, and so certain that I was destined to go the same way, that the next morning he hid my flying boots. Not surprisingly, that little ruse failed to achieve its objective. And so he took to substituting for my dawn cup of tea a large glass of zabaglione, made with about four times the specified amount of Marsala. Perhaps his objective was to keep me from flying by rendering me incapable of doing so. Fortunately for me, the reverse effect was achieved; I rather took to the stuff, which got me off to a good start every day.

On 27 April and the morning of 28 April bad weather curtailed our activities. On the afternoon of the 28 April cloud lifted; 601 and 145 Squadrons spotted some two hundred German vehicles nose-to-tail north of Vicenza. They claimed fifty destroyed in flames and as many more seriously damaged. Seven of our Spitfires were hit.

On 29 April Venice was captured.

The next day – the last of that blood-stained month – we were operating at the limit of our range, without bombs. Many vehicles were claimed destroyed by strafing. Four Spitfires were lost, one of their pilots killed, the others baling out or crash-landing. That was the last big day. On 2 May there were only three missions, involving four aircraft each, and they had instructions simply to report enemy movement, not to attack.

Five days later the Wing Log records: 'We no longer carry out operations from this date. In other words the war in Italy has ceased.'

And I was still alive.

On the eve of VE Day 244 Wing landed at Treviso, a well-equipped airfield, with comfortable quarters, a few miles north of Venice. Once again, my luck was in. Exceptionally experienced in the hard business of aerial warfare I may have been, but I had a lot of catching up to do in other directions. And I could hardly have been better placed for the purpose, with the glorious city of Venice on my doorstep (I was put in command of the Lido airfield, as well as of Treviso, and was provided with a comfortable, fully staffed, villa there for the entertainment of visiting VIP's who were numerous); with Cortina d'Ampezzo less than two hours drive to the north, in the Dolomite mountains; and with Rome a mere sixty minutes away by Spitfire.

Slowly, I unwound. And, from this fortunate vantage point, I spent two wonderful summers and one winter exploring and discovering some of the pleasures of life which I had been missing. And so I was able to enjoy a cheerful intermezzo before buckling down, in the hard, cold austerity-bound London winter of 1946/47, to the serious business of earning a living away from the cockpit of a Spitfire.

Letters

In 1940, my brother John and I corresponded regularly with our parents and with each other. When my mother died, in 1979, a bundle of these letters was found among her possessions. It is, perhaps, a revealing fact that, although I wrote to her almost weekly throughout the war years, the last letter she kept was one written on 1 April 1941, just after she and my father had been formally told that John must be presumed dead. I think that I did not, then or later, fully understand what a devastating effect that final pronouncement had on her.

The first letters reproduced were written at the time of Dunkirk. I have referred, in Chapter 1, to my own letter, describing to my mother my first engagement with the enemy, and have indicated that it was a somewhat colourful account. Certainly, it brings a blush to the cheek. And certainly it was not a missive calculated to soothe a mother's fears. But it must be remembered that it was written by a boy who, only six years before, had still been at prep school.

I think that the letters, though of absolutely no historical significance, may be of some interest as a reflection of the family background against which two young brothers played their parts as fighter pilots during that dramatic and blood-stained summer and autumn.

1.6.40
R A F Rochford
Southend

Dear Mummy,

On the second patrol we did, seven of us ran our necks into 30 or 40 Messerschmitt 109 single seater fighters. We were hopelessly outnumbered – every time I manoeuvred to get a shot 2 or 3 would dive down from behind and shoot me up, so that it was almost impossible to get a fair shot. The CO was out of it almost right away, with a punctured petrol tank and instrument panel; Dick Hellyer got shot and put down on the beach at Dunkirk; George Moberley got one with his first burst, but he was put out of action almost immediately; a sergeant pilot got a bullet through the scalp, but got back here and is perfectly O.K. now; 'Scottie' peppered one very badly, but was soon out of it with a couple of cannon shells through his main plane; Ken Holden got one, and ran out of ammunition. In the end, after about 8 minutes, I was the only Spitfire left, and was just about to push off when a couple which attacked me from the beam split up and gave me a chance; by turning sharply into them I managed to split them up and get onto the tail of one; he dived like a rocket, and I pumped a lot of lead into his starboard mainplane, where he keeps his petrol. We both disappeared into a cloud at about 400 mph, and that seemed to me an excellent moment to push off home! A moral victory, we thought, considering the odds; Dick is back in England and he'll soon be flying again.

In the 'Dawn Patrol' this morning (very much à la cinema show) Red Section ran into something before my section, which was last off, arrived on the scene. When we got back Ken Holden and Jack Bell were missing, but Ken rang up from a farm house where he had force landed, said that he had got 2 more 109s, and that he thought Jack was O.K. Ken is becoming the Ace; he is a grand, tough, burly Doncaster man with a broad Yorkshire accent and an unfailing sense of humour.

Joe Dawson put down here 2 days ago after 609's first patrol out there; he and John and another man had come back in thick weather without much petrol, and he was laughing heartily because John had had to land on the playing field at Frinton, trickling slowly through the fence at the far end! John and Dawson are the two great friends and rivals in 609; they started at the same time, were the two promising pupils and great competitors – whatever Joe did John had to do one better, and vice versa; hence the amusement about the playing field fence!

We like this place very much really, but probably won't be here long; after another week or so we'll probably be relieved by another squadron, and go back to Leconfield ourselves to get new aircraft and train 2 or 3 more pilots before our next crack. I wouldn't worry your head about John or me, we are on a good job really: I am quite certain that our Spitfires are the finest machines flying over Belgium now, and they seem to be able to fly even when they're only hanging together by a few wires.

The best of luck to Daddy: I hope he gets some 'game' to shoot at too!

Lots of love

Hughie

R.A.F. Northolt
Ruislip
Middlesex

2.6.40

Dearest Kanga,

Just a line to say that for the time being things have quietened down and that I'm still alive and kicking and in one piece. The last 3 days have been very hectic – patrolling Dunkirk most of the time. After one patrol I force-landed with o gallons of petrol on a cricket field in the middle of Frinton on Sea. The poor aeroplane came to rest on its wheels and its prop in a kneeling attitude in some iron railings. No one was hurt. The day before yesterday we had a real free- for-all with a lot of Huns. I got a Heinkel down in flames amd damaged a Junkers. Altogether in that engagement we got 6 confirmed and 10 possibles. Yesterday I loosed off all my rounds at another Heinkel, but the wretch refused to come down. However, it jettisoned its bomb and fled. So did I.

I'm afraid we've had some losses, and may go up North for a bit to recuperate and make up our numbers.

You never enclosed Hughie's letter, but I knew all along where he was and hear that he scored over a Heinkel some days ago. Nice work –

It's v. hot and close here – spent a busy morning this morning checking sights and guns and going round my aeroplane (fortunately it was a borrowed one which I parked at Frinton) and am released this afternoon. Like Uncle Bill, 'my sole idea of fun is sitting snoozing in the sun'. I may ring up Harold and if he's free invite myself to supper tonight. I think he'd like to hear my news.

Love to all

John

RAF Northolt
Ruislip
Middlesex

9.6.40

Dearest Kanga-

Nothing to report here for the last week. They haven't given us any 'let-up', but on the other hand they haven't sent us on any more foreign excursions. I rang Rochford the day before yesterday, but was too late: H. had gone back to Leconfield the day before. His CO was over here the other day.

Both H and George Moberley seem to have done well. One of our chaps who had been given out as missing – a Sergeant Pilot – reappeared in hospital near here yesterday. He'd been shot down in the sea and was picked up by a trawler. Another of our chaps has been awarded the DFC. Unfortunately he is not to receive it: he's been missing for a week and it looks as though, at best, he's a prisoner of the Nazis.

I was so glad to hear about Robert and Miles Howard. Having seen the thing at first hand, I still don't understand how they managed to get all those men out.

I expect you're fed up with war news; but there isn't much else to tell you about. I've seen Margaret Rawlings once or twice – on the rare occasions when I have been able to get off the station. Since 'The House in the Square' came off (I managed to see the last performance of it, and we had a very gay party at the Hungaria afterwards) she has 'given up' the stage and now drives a large Crossley round London and district on missions connected with Government contracts. She wears a black coat and skirt and a large chauffeur's cap which gives her an 'apache' appearance – and is very proud of the fact that by taking this job she has released 2 skilled mechanics for work on aeroplanes. Stephen Beaumont, by the way, is fighting fit.

Like a stupid fool I've been and gone and forgotten Pooh's birthday. Do please send this letter on to her with my apologies and greetings to Patrick. I hope Ann is in good form and not quarrelling with the heat.

Well, au revoir. I hope that before too long I'll get a few day's leave. But I don't know whether I will – it's only a hope.

Love to all

John

R.A.F. Kenley
Whyteleafe
20.8.40 Surrey

Dear Mummy,

This is to let you know that we have arrived safe and well, and to tell you what sort of place this is. Geographically it is just south of London, near Croydon (no information to the Bosche there – he's found it alright!). It is a very pleasant station, flat out for the pilots, 'vide' a huge, hot and altogether excellent breakfast which, to our immense surprise, was awaiting us when we got up at 4.0 a.m today; I've never found anything like that at any other station – what's more it was equally huge, hot, excellent and fresh when I went in for another dose at 9 o'clock!

Nothing much has been doing today, which was quite welcome as giving us a chance to find our feet and settle down. Our flying quarters are a bit odd at the moment, as, the day before we arrived, a Bosche, together with bombs, effected an unwilling landing which rather devastated everything, including our hut. However one Hun is considered to be worth quite a lot of huts and discomfort, and things are getting straight again pretty quickly.

With luck and good management I should be able to make contact with John now that we are both in the same part of the world – that is if he hasn't moved; I haven't heard from him for some time, and I don't know exactly where he is. I have written him today, suggesting a well coordinated movement to the same place. There is rather a distinguished old Aysgarthian here, one Squadron Leader Kyall, who has recently been awarded both the DSO and the DFC. He was there just before me I think. So, with Dick Hellyer, one of our Flight Commanders, at least 3 of that school are here!

All love to you: I will write as often as possible,

Yours, Hughie

Kent & Canterbury Hospital
Canterbury

23.8.40

Dearest Mummy,

This is a quick letter to let you know that all is well with me, despite the ominous change of address. I had to jump out yesterday evening when two explosive shells at 12000 feet put my machine out of action and set fire to it. Please don't worry, as I really had quite an enjoyable parachute descent! Moreover my injuries are very slight; the machine was blown about quite a bit, and the jar dislocated my left shoulder, and a lot of little splinters got into my leg; but there is only one puncture worth speaking of, the rest being mere scratches: I can even walk about. The shoulder has been put back, and is now quite comfortable in a sling. With a bit of luck I should be visiting Dale Cottage again before long. The Doctor tells me that I can't possibly fly again for 6 weeks.

I saw John yesterday, when he flew over for luncheon: it was a surprise, and a very pleasing one. He looked surprisingly well, and was full of life.

I have also seen, since going to Kenley (a) The Duke of Kent, who was over at our dispersal when we came in from patrol, and shook hands all round, (b) Winston Churchill, who also came to see us at our dispersal.

This is an excellent hospital. I am in a private ward with another Fighter pilot, who has had a bullet through his head, though I must say that he looks remarkably well on it! I will let you know when I am likely to be out — probably in one or two days I should think — and what will then become of me.

Lots of love, and don't worry,

Hughie

Saturday 24th Aug. 40

Kent & Canterbury Hospital

Dear Mummy,

It was grand to get a letter from you this morning, and I hasten to reply, though you will have received my first letter by now. The doctor told me this morning that I should probably get away from here by the beginning of next week, go to Kenley for a board, and then home till I'm fit enough to fly again. The shoulder is quite comfortable, though of course I can't use that arm at all;

(there is a good raid going on over here right now, which makes writing a bit disjointed).

As for the leg, there's nothing much wrong with that, except for a lot of tiny little holes full of aluminium, which I intend to collect for the national cause!

George came over yesterday evening and spent the night at Canterbury; he tried to ring you up, but couldn't get through. He tells me that I was shot down by the 109s, which came out of the sun and attacked our rear section. I was the only one they got out of action before George saw them, and swung the section round. The other two in the section routed the 109s while George flew down to protect my tail, as they continued to attack as I fell; he followed for about 7000', and as I still hadn't bailed out, and was burning merrily, he gave me up and rejoined the Squadron. Actually I was having some trouble getting clear, and didn't get out till about 800 or 1000 feet; when at last I managed it I really quite enjoyed the jump, which showed me the world at an entirely new angle!

I am going to get up this afternoon to sit in the garden, it being a glorious and sunny day. The sirens have been going all day, with guns and bombs in the distance, and occasional machine gun bursts overhead. I have never seen it from the ground before – it all feels very strange and out of place viewed from this peaceful and sleepy old town.

I shall hope to see you soon. If George can get some leave, which he is overdue for, he is going to drive me up – would there be room for us both in the house?

> *All love,*
>
> *Hughie*

> *Royal Air Force*
> *Middle Wallop*
> *Hants.*
> *Tel: Middle Wallop 344 (ex.24)*

25.8.40

My dear Hughie,

Very sorry indeed to hear that a 109 – or rather 12 of them – inflicted grievous bodily harm on you over Dover two days ago. Mummy sent me a wire yesterday, and you were mentioned as wounded in a 11 Group Intell. Summary this morning. I haven't heard any details, but I do hope the damage isn't too bad. Write and tell me about it as soon as you're well enough to do so. Anyway you'll get a nice spell of sick leave, which I rather envy you.

The 109s nearly made hay with us over I of W yesterday. They sent us off too late to do anything about the bombing of Portsmouth and too low to do anything about the myriads of 109s who were hovering around the scene and who, when they saw poor old 609 painfully clambering into the sun, came down on us. The result was that one of our machines was shot to hell, two more damaged and not one of us succeeded in firing a round. I was reduced to the last resort of a harassed pilot — spinning. It was most humiliating. But fortunately we didn't lose any pilots. Before long we'll get our revenge, I hope.

Do write as soon as you can and let me know all about it.

Love,

John

Officers Mess　　　　　　　　　　　　　　　　　*R.A.F. Station*
Tel: Middle Wallop 344 (Ex.24)　　　　　　　　*Middle Walop*
29.8.40　　　　　　　　　　　　　　　　　　　*Hants*

Dearest Kanga,

Very many thanks for your letter: so glad to hear that Hughie has got a fairly long leave. I still haven't heard any detailed account of just what happened, or how, or what injuries he suffered. Do tell him to write and give me all details.

It almost looks as though — touch wood — August may peter out fairly quietly — only 2½ more days to go now. Our last engagement was on Sunday, when we got in among a gaggle of ME 110 bombers and shot them down in considerable numbers. 609 shot down 13 of them and altogether 15 crashed on land near Warmwell, the aerodrome in Dorset which they were trying to bomb and where 609 has spent quite a lot of time on and off. I felt very pleased about it because (a) we lost no one and (b) I was leading that time and managed to bring the chaps in against the leaders of the Jerry formation which broke up the raid completely. It was a most gratifying sight.

The Dentist was then doing a Blitzkreig on me: he found my teeth full of holes, and one, near the back of my mouth, had to come out. Besides, he's stopping 8 holes and has 4 more to do. He's quite a good dentist and used to be Nyns' assistant in Doncaster and I get his services for nothing. He says that repairing my teeth is more a feat of engineering than dentistry owing to the large amount of fillings etc which they already contain! I think its pretty definite that I'll be getting leave from 8 to 12 Sept.

Love,

John

End Sept 1940   *As from*
                                                  Kirton in Lindsay
                                                          Tuesday

Dear Mummy,

This is the best I could do now in the line of ink and paper. I am sitting in a hut on the edge of a field in Cambridgeshire, where we come at crack of dawn every day. We patrol Piccadilly as a Wing of 5 Squadrons, superbly led by S/ Ldr Bader, of Wooden legs fame. It's the best thing I've seen since this war started; instead of being almost invariably practically alone among many huns, we are a large, concentrated and formidable looking formation. We have done several patrols in the last six days, since we joined the Wing, and rarely get near the Hun, as he turns and belts off home as soon as he smells us. As I say, its the most heartening thing I've seen since this war began; at last someone has got things organised.

John told me about his leave and I have asked him to let me know when he will be at home. It won't be easy to get time off, as we are short of experienced pilots and so far I have been on every time.

Things look up every day in the Squadron. The more I see of the CO and Mackfie, my Flight Commander, the more I think of them. All we need now are a few bloodless victories to boost up morale, and there's no doubt that we are in the right place to get some.

John wrote urging me to get posted to 609; much as I would like to be with him – and I know that you would like us to be together – there is no question at all of my pushing off just now. The circumstances just don't leave any two ways about it, and I am sure that, were John in my position, he would feel the same.

For Heaven's sake don't worry at all about me now. We're on a grand job, and very lucky and honoured to be included in this Wing, which I think is keeping Jerry off more than anything else: we are in company with some pretty crack Squadrons.

                                          Yours with love,

                                                          Hughie

P.S. I said all this in a letter about 2 days ago, but I'm pretty certain I didn't post it!

Royal Air Force
Middle Wallop
Nr Stockbridge
Hants
Tel: Middle Wallop 344 (ex.24)

6.10.40

Dearest Kanga,

Good news! When I got back here I found I'd got B Flight and have been given the DFC. The Station Commander had the ribbon tucked away in a drawer in his desk and produced it immediately, so I am now wearing it and feeling rather a fool. Not that it's worth much these days. Our CO, S/Lr Darley, has been sent to Exeter as Station Commander and we have a new CO called Robinson, no relation of H's ex-CO. I like him on first impression. He's young, energetic and rather off-hand. Also he's a crack pilot. Darley's got the DSO and McArthur, now in hospital, the DFC. So 609 is getting some recognition at last.

The weather's been completely dud today, so we've just sat on the ground, but I hope to get everything organised in the next few days.

Very many thanks for giving me such a wonderful leave. You've noticed how much good it did me: I completely forgot about the war. For hours on end. But, I'm afraid Daddy will have rather a stiff petrol bill.

I've been put up for promotion to Flight Lieut. but it probably won't come through for a long time.

Much love,

John

Photographic Acknowledgements

The author and publishers are grateful to the following for their kind permission to reproduce copyright photographs, identified by caption:

Hulton Picture Library: Second anniversary of the Battle of Britain; Spitfires taking off for convoy patrol; Dunkirk – the German target; remains of a Heinkel 111; remains of a Spitfire; bus stop.

Imperial War Museum: Baltimore light bombers in action near Salerno; desert air force interdiction; 56 Squadron at Duxford; 'Stimmie' Stimpson with a typhoon of 56 Squadron.

Popperfoto: John Grandy; Douglas Bader; pierced steel planking; Italian bomber-eye view of Hal Far airfield; Colin Macfie.

Photographic Acknowledgements

Index